The Young Child as Scientist

A Constructivist Approach to Early Childhood Science Education

Christine Chaillé
and
Lory Britain

University of Oregon

HarperCollins*Publishers*

Executive Editor: Chris Jennison
Project Editor: Steven Pisano
Art Direction: Teresa Delgado
Cover Design: Delgado Design, Inc.
Cover Photo: Fredric Petters Photography
Production Administrator: Beth Maglione
Compositor: BookMasters, Inc.
Printer/Binder: R. R. Donnelley & Sons Company
Cover Printer: The Lehigh Press, Inc.

The Young Child As Scientist: A Constructivist Approach to Early Childhood Science Education

Library of Congress Cataloging-in-Publication Data

Chaillé, Christine.
    The young child as scientist : a constructive approach to early
childhood science education / Christine Chaillé, Lory Britain.
        p.   cm.
    Includes bibliographical references and index.
    ISBN 0–673–38857–3
    1. Science—Study and teaching.   2. Science—Study and teaching
(Elementary) 3. Science teachers.   4. Constructivism (Education)
I. Britain, Lory.   II. Title.
Q181.C414 1991
372.3′5—dc20   90–27773
                                                           CIP

91 92 93 94   9 8 7 6 5 4 3 2 1

Contents

Preface

A book title needs to be carefully considered and selected. A title can imply philosophy, content, and promise to the potential reader. Our title, *The Young Child As Scientist: A Constructivist Approach to Early Childhood Science Education*, is intended to imply that the child is *like* a scientist, rather than only a learner of science content. This "child as scientist" analogy simply means that many of the traits associated with a scientist—experimenting, curiosity, creativity, theory-testing—are also typical of young children. For many of us, this conjures up two parallel images—the stereotypical scientist intently experimenting in the laboratory, concocting exotic formulas, and the young child purposefully mixing sand, mud, bits of rock, and water, and pouring the mixture through a sieve.

Consider the implications of this comparison for teachers. If you, as a teacher, understand the importance of supporting young children in their development, then you can also understand the nature of supporting children as scientists. Thus, another title for the book might have been *The Understanding (or Aware) Teacher as a Science Facilitator.*

For many teachers, who conceptualize science as specific information to be "taught" to young children, science has been relegated to having children follow directions from a science textbook for twenty minutes, twice a week, or to only when the "science teacher" comes once a week, or to the presence of a "discovery table" on which sit increasingly dusty seed pods and seashells. Or perhaps science in the curriculum has somehow been pushed to the bottom of the priority list because the teacher doesn't feel qualified to "teach" science.

But the constructivist version of the child as scientist frees the teacher from those constraining situations and opens up new worlds for the teacher to explore as facilitator of children's scientific explorations. Because constructivism focuses on the child's own perspective on the world, on the child's self-directed interactions with both the physical and social worlds, it is familiar territory for a good teacher. And, for the new teacher, it is territory that needs to be explored.

It is this territory that we have set out to explore. We have tried to describe how to create a learning environment that supports theory building by taking three traditional content areas of science—chemistry, physics, and biology—and "translating" them into developmentally appropriate constructivist educational practice. We have also given you some theoretical ideas that are not couched in theoretical terms, and provided you with lots of scenarios of young children engaged in science.

Some think of "constructivism" as a *new* theoretical perspective. The constructivism of this book is not new, it is Piaget's constructivism, the term he and many others have used to describe how children learn (see DeVries and Kohlberg, 1987). We hope this book contributes an accessible and personal view of what constructivism means in implementing science education for young children. We envision this book being used in college classes, probably as a supplement to broad early childhood curriculum or elementary science education textbooks. The book is also written for practicing teachers interested in constructivism and/or different approaches to science, and could be used in inservice programs. Our audience is anyone who is curious about how children learn, anyone who wants to change to be more constructivist, and anyone who wants to know what it could look like.

Our application of constructivism to the teaching of science education has deep roots and many participants, many of whom have been associated with the Early Childhood Center, our laboratory school at the University of Oregon Division of Teacher Education. Among those whose influence pervades our work are Christine Pappas and Pamela Perfumo who, with Christine Chaillé, founded the Center in 1982. Barbara Littman's influence, first as a student and then as a curriculum consultant, on the curriculum model and many of the activities developed was also important. The students in courses (particularly the course Constructivist Theory and Its Application) whose ideas provoked our thinking are too numerous to name. We can, however, thank others who had direct input into the book. These include Joanne Wiser and Michelle Brenner of Irvington School, Portland Oregon, and their principal Pam Shelley, for graciously permitting us to photograph children in their appropriately constructivist prekindergarten and second grade classrooms; Stephen Lafer of the University of Nevada, Reno, and Janice Jipson, our colleague at the University of Oregon, for reading early drafts and giving great advice; our secretaries, Linda Zimmerman who edited various drafts, and Linda Kelm, who helped in many other ways; the reviewers, Bess-Gene Holt, James Shymansky, and Elaine Surbeck who raised critical issues and whose comments helped us refine our ideas and our writing; Eric Leonidas and Steve Pisano, project editors at Harper-Collins, who were clear and helpful throughout the production process; Chris Jennison, for his intelligent and supportive role as acquisitions editor; and, finally, our families, particularly Greg, Jima, Peter, Adrienne, Sierra, and Dylan, for their support and inspiration.

Christine Chaillé
Lory Britain

PART ONE

The Constructivist Perspective

Chapter
1

The Child as Theory Builder

The focus of this book is children—how they experience the world, interact with each other, pose questions and problems, and, in the process, construct knowledge. Everything that we have to say reflects this focus on the child: thoughts about how we should organize materials and activities, what our role as facilitators should be, and the curriculum model we present in this book. As we have looked to children and attempted to understand how they think and learn, we see striking similarities between children and scientists.

For example; consider Adrienne's experimentations with a swinging pendulum:

■ Adrienne slowly approaches a swinging pendulum which has four paint-covered brushes attached to it vertically. When the pendulum is still, the brushes just touch the large piece of cardboard set on the floor under the pendulum. Adrienne stops and stares at it, her eyes moving from the brushes up to the rope and to the hook from which it hangs. The brushes make a mark on the cardboard until they stop making contact with it. After Adrienne inspects the mark, she runs around to the opposite side and gives the brush pendulum a shove in the opposite direction. It makes a mark almost identical in length to the first. Adrienne tries shoving harder and then, dodging out of the way, bends down to inspect the mark, which is the same length as the first two. Next she tries holding the pendulum brush with her hand,

This activity with a swinging paint pendulum provides the children with opportunities to consider many variables and relationships—for example, the incline of the cardboard, the path of the pendulum, the force of their push on the pendulum, and the paint marks on the paper.

trying to make a brushstroke on a different part of the cardboard. Unable to do this, Adrienne tries to lift the large piece of cardboard up on one end as she swings the brush pendulum. She is unable to hold it up. She puts it down and runs to the block area for three large blocks and puts them under one end of the cardboard, propping it up on an incline. Adrienne then runs to the other side and shoves the brush pendulum quite hard. As it swings it knocks down the cardboard and blocks. Billy, who has been watching quietly all this time, comes closer and inspects to see if there is a mark on the cardboard. "Look, you made a dot here," he says to Adrienne as he points to the latest mark. He picks up the end of the cardboard that had been held up by the blocks. Laughing, Adrienne says "Let's make dots!" Billy laughs, saying "Yeah, let's make dots," as he moves the cardboard around to make contact with the swinging pendulum brush.

Adrienne is engaged in lots of experimentation. She is exploring the relationship between her shoving motion and the marks on the paper. She is also experimenting with the relationship between the brush and the height and position of the cardboard on which she is making marks. Billy, picking up on Adrienne's experimentation, focuses his experimentation on what happens when he moves the cardboard, thus reversing Adrienne's action, which dwells on the brush.

Scientists, seeking to understand an unknown world by way of experiment, are continually doing the same things that we see Adrienne and Billy doing: having insights, asking questions, solving problems, trying out new ideas. Scientists, like children, do not simply apply systematic methods to answer predetermined questions. Scientists—filled with wonder and curiosity—are constantly puzzling, testing, and probing ideas, just like children.

The constructivist perspective that we present in this book focuses on what we know about this process of knowledge construction that is shared by both children and scientists. In thinking of all that we know about young children and how they think and learn, we have found there are four characteristics that are of utmost importance for educators and that underlie much of what we will present in this book. These characteristics are as follows:

1. Young children are theory builders.
2. Young children need to build a foundation of physical knowledge.
3. As they mature, young children become increasingly autonomous and independent, both intellectually and morally.
4. Young children are social beings.

CHILDREN AS THEORY BUILDERS

Young children are mentally and physically active, continually engaged in the process of building theories in all domains of knowledge: language, reading, mathematics, art, music, and also science. To draw an example from the physical world, the very young child pushes a round object and it rolls. "All objects roll," is the theory. The child pushes a square object; it slides, but it doesn't roll! This contradicts the theory and any prediction the child might have made. Now the theory must be modified, becoming, "Round objects roll, cubes slide." The development of increasingly complex theories would be based on experiences with all kinds of objects, lots of different surfaces on which to place them, and varying inclines. These experiences stimulate the building of new theories. But this process also depends on the child's confidence and flexibility, since children, like scientists, must be open to new ways of seeing things; they need to give up the old ways. The new information—"cubes don't roll, they slide,"—also means that the child's original prediction was "in error." The process of theory building is full of error, conflict, and contradiction. It has to be!

Consider as an example the 4-year-old who is building a complex structure, incorporating a bridge over two upright blocks. First she tries placing two blocks horizontally, one on each upright block, but there is still a gap. When she moves the blocks over to fill the gap, they fall. She tries two different, longer blocks; they still fall. Then she gets one very long block, places it over the top, and a bridge is built. In this example we can try to identify the hypotheses and predictions the child was making at each step. "Maybe longer blocks will bridge the gap." And "What will happen if I try just one very long block?"

We can also see that the process of theory building changes with age, as children become able to understand things in different ways and as their theories become more complex. For example, a 3-year old will probably hit a target with a swinging pendulum ball by grasping the rope just above the ball in order to direct its movement. An older child would be able to use his or her understanding about the path of the pendulum to release the ball at just the right point to knock down the target.

Science educators are becoming increasingly aware of the sometimes complex and varying theories that children develop about natural phenomena. The work of Driver (1983) and Osborne and Freyberg (1985) has stimulated many to examine what has been termed "children's science." This is yet another indicator of children's natural propensity to try to make sense of their experiences.

Our approach to science emphasizes theory building, which is an important activity of young children.

Focus on Physical Knowledge

Before considering why young children need a strong foundation in constructing physical knowledge and why this might be a curricular focus, we must examine the different types of knowledge that children acquire. Piaget (1970) describes three types: physical, logico-mathematical, and social knowledge. Physical knowledge involves the understanding of the physical world—how objects and materials behave as a result of their characteristics and attributes. An example of physical knowledge is the one given above, where the child learns about the movement of balls and cubes. This knowledge is "out there" in the world; it is empirical, and it lays the foundation for later, more abstract thought that involves understanding things that are not empirical.

The second type, logico-mathematical knowledge is not observable or empirical. It involves the construction of knowledge about relationships between objects through the use of comparison and seriation. Determination of size—whether something is "big" or "little"—depends on what an object is being compared with. Relative size is attributed to objects by you, the observer, depending upon what you are placing into the relationship. For example, a chair is small compared with a table, but it is big compared with a marble.

The third type, social knowledge, is confusing, partly because of the name Piaget gave it. It is not knowledge *about* the social world or social interaction; rather, it is knowledge that can only be transmitted socially, such as customs, particular names, and labels for things—anything that is clearly culturally determined and therefore arbitrary. There is nothing in a tree, for example, that implies it has to called "tree," which is why it can be called different things in different languages. Because social knowledge is arbitrary and culturally determined, children *cannot* construct it. For example, there is no way that a child can construct the name "tree" by touching the tree, poking at it, or looking at it. The child must be given that social knowledge (the word "tree" in this case) by another person. Social arbitrary knowledge must be transmitted, one way or another, from the culture to the child through direct adult-child teaching, books, or other media.

All kinds of knowledge are important, including social knowledge. In order to communicate, for example, we *need* to know the labels for things in our culture. However, because of the way young children learn, we must be cautious about placing too much emphasis on social arbitrary knowledge. Children need ample opportunities to construct, through their own mental and physical activity, physical and logico-mathematical knowledge. In the classrooms, we often overemphasize social arbitrary knowledge at the expense of those types of knowledge that are "constructable" by the child.

All three types of knowledge are also interrelated. Therefore it is difficult to pinpoint the type of knowledge that is involved when you look at a child's *behavior* in a particular instance. For example, Adrienne's experimenting, described above, contains elements of all three types of knowledge; she is constructing physical knowledge related to the height of the cardboard and the effect of the pendulum swing, she is constructing the logico-mathematical knowledge involved in comparing marks of varying length, and she is labeling the marks "dots" when they look a certain way. You can see, however, that the pendulum painting activity emphasizes knowledge that is constructed by the child—physical and logico-mathematical knowledge.

Educational experiences for young children, then, should emphasize the *construction* of knowledge, not its transmission. This distinction will be developed further in later sections.

Increasing Autonomy

As they grow young children progress from heteronomous dependence upon adults to increasing independence and autonomy. Put simply, "heteronomy" means being subject to external laws or domination, while autonomy is self-government. The infant is, of necessity, dependent on an adult to satisfy his or her needs, and thus is, for the most part, externally controlled. This is not to say that infants do not contribute to the complex interactions between them and their caregivers. But as children grow, they become *increasingly* competent and able to direct themselves. Children's sense of self is at least partly dependent on seeing themselves as capable and independent beings. By giving children opportunities to be independent, to make choices where appropriate, and to rely on themselves rather than on adults for materials and direction, we can greatly influence their sense of self, their confidence and sense of mastery, and their awareness of themselves as active constructors of knowledge—all aspects of being autonomous.

Yet many of the ways in which we interact with young children actually *foster* dependence on adults, making the child see the adult as the source of knowledge. Constance Kamii (1985) discusses how many of our methods of education work against the development of children's autonomy. In expecting blind adherence to arbitrary rules, using authority as the source of "truth," and thinking of knowledge as something to be "transmitted" from adult to the child, our educational settings will tend to foster heteronomy.

In contrast, when children are given opportunities to solve problems and to experiment in self-directed contexts, we facilitate their development of autonomy. Adrienne, for example, was faced with many possibilities in confronting the pendulum paintbrushes. Having made her own choice, she then has to face the results of her own actions and must try to figure out why the paintbrushes did or did not make a mark. In this context, she felt free to run over and get blocks on which to raise the cardboard, one of the problems that presented itself, and she effectively solved it. The lessons she learned here went beyond learning about pendulum arcs and have a lot to do with her perception of herself as an independent problem solver.

Children's needs to develop both intellectual and moral autonomy should be taken into account when we develop curricula and determine

Social interaction actively contributes to children's theory building. These two girls collaborated and decided to see what would happen if they put wet tissue paper on the windowsill.

how adults should and should not interact with young children. One of our goals should be to facilitate the development of autonomy.

Children as Social Beings

Young children are social beings, constructing an understanding of their social world and interacting with others as they learn about the world around them. Warm and supportive relationships with adults and ample opportunities to interact with peers are essential if children are to develop an understanding of the increasingly complex society into which they are growing. Good environments for young children permit, encourage, and even necessitate interaction with others, from simple communicative interaction to the complex negotiation of conflicts.

But social interaction is important not only because it is a part of life, but also because it actively contributes to children's theory building. When children interact with other children, they are confronted with different ways of seeing—different perspectives, different solutions to problems, different answers to questions. Research such as that of Perret-Clermont (1980) has shown that children respond to others' perspectives differently when their source is another child as opposed to an adult or some other source of authority, such as a book. Children are more open to the conflicts that differing perspectives may arouse in them, and then are likely to make the effort to coordinate their own perspective with others in such social contexts. This resolution of conflicting perspectives can, in turn, lead to reflection and intellectual growth.

In addition to the growth-enhancing intellectual conflict generated by interactions with others, children may also learn by imitation. This is a powerful effect of sharing an environment with peers. Children who are actively experimenting are always in search of "good ideas" and will glean them from the actions of others when they can. And such activity is not limited to the imitation of peers; adults who themselves are inquiring problem solvers serve as good models and can stimulate children to approach problems in new ways. For example, a group of preschoolers, finding the skeleton of what looks like a dog, can learn a lot from the adult who says, "I wonder what we could do to find out what this is?" They would learn something quite different from the adult who says "That's the skeleton of a dog." Consider, as another example, a situation in which a paper airplane is stuck in a tree. The adult says to the child, "I wonder how we can get it out of the tree?" This is a more stimulating approach than the suggestion to "go find a stick and knock the branch with it." In the former cases, the adults are providing models for problem solving and inquiry rather than acting as sources of information or answers.

As with the development of autonomy, educational settings all too often deny the importance of the social context of learning, and overemphasize the interaction between teacher and group. Our aim is to encourage children to interact and construct knowledge in a social context as well as to construct understanding of the social world. We do this by providing the physical contexts in which children *can* interact with each other; offering support and encouragement for negotiation, problem solving, and cooperation; and removing the teacher from center stage.

GENERAL IMPLICATIONS

Keeping in mind what we know about young children—that they are theory builders, need to focus on physical knowledge, and are social beings who need support in order to develop autonomy, we can make several generalizations that constitute the essence of the constructivist perspective on education.

Our Conception of the Learner

The learner is actively constructing knowledge rather than passively taking in information. Learners come to the educational setting with many different experiences, ideas, and approaches to learning. Learners do not *acquire* knowledge that is transmitted to them; rather, they *construct* knowledge through their intellectual activity and make it their own.

The constructivist conception of the learner acknowledges the different, more or less complex prior understandings and cultural values that the learner brings to the educational setting. There is also an acknowledgement that transmission by the teacher does not of itself lead directly to learning. Ultimate responsibility for learning lies with the learner, and even the best presentation of information may not be "received" and processed.

Imagine a kindergarten teacher who "presents" to children rules about which objects sink and which float and then asks whether a piece of aluminum foil will sink or float. Obviously this teacher is ignoring the fact that 5-year-old children have already had many chances to construct theories about objects and how they act in water. That group of 5-year-olds may include some children with very clear ideas about sinking and floating, others with "strange" theories about sinking and floating, others with incomplete theories, and yet others who haven't a clue. Each of these children will "receive" the teacher's presentation differently, depending on his or her prior knowledge and current constructions about sinking and floating.

Learners come to the educational setting with diverse ideas, values, and approaches, and this needs to be more than acknowledged; this diversity must be welcomed.

Our Conception of the Educational Setting

The educational setting provides contexts in which the learner constructs knowledge—contexts that encourage self-direction, experimentation, problem solving, and social interaction. These contexts must be supporting, stimulating, and facilitating, helping the child to construct knowledge. Educational contexts must allow for flexibility across space, time, and curricular areas.

For most preschool, day-care, and kindergarten environments, this may not sound very revolutionary, since good preschool/kindergarten practice has always incorporated provisions for play, social interaction, and self-direction to some degree. However, we do need a more radical reconceptualization of the primary school classroom. At the primary level, one must envision classrooms that may look somewhat different from the way they are traditionally pictured. Physically, the classroom setting may look different, with places where materials are accessible and used without teacher direction, where traffic patterns let children move easily from place to place, and where children are encouraged to interact with each other, most commonly in small groups. There may be more noise and activity and certainly more diversity in what children are doing at any one time. And, critically important, there would be less segmentation of curricular areas. Thus it might be difficult to pinpoint the subject-matter area that the whole class and individual children are working on; many different and integrated subject matters would be represented. Children would more likely be working on projects that incorporated numerous possibilities for engagement. The "project approach," which has been promoted by Katz and Chard (1989) and Gamberg, Kwak, Hutchings, and Altheim (1988), is thus very consistent with the approach of this book and can help teachers to conceptualize the constructivist educational setting.

■ Envision a second-grade classroom in which several children in one corner of the room are constructing a house made of a large cardboard box. They are trying to attach fronds of palm leaves and are faced with many problems in doing this, in constructing an archway across the front, and in covering the roof portion. These children came up with the idea of making their own house after reading a book about a tribe of people whose houses were made of large palm leaves. From the reading of this book and the children's idea of building a house grew a rich activity that integrated cultural studies, math, science, and literacy in the context of rich cooperative group interaction. For example, the children became interested in reading and hearing more about how people build houses in other cultures. They had to measure and estimate lengths of the various materials they used to construct their "houses." Such principles as weight, balance, and the properties of inclines had to be considered. And the children decided to make signs and dictate their descriptions of the house-building process.

Notice that when such integrated, child-initiated projects occur, the teacher has to allow flexibility in the class schedule and the use of room space. There is tolerance and encouragement of lively discussion and a focus on children's self-initiated activity—the cornerstone of the constructivist classroom. Because this can be a formidable task, particularly challenging to the new teacher, we will try, in subsequent chapters, to lead you through the elements of designing the physical and social environment that can help in this task. Most important, however, is to rethink the role of teacher in the constructivist classroom.

Our Conception of the Role of the Teacher

The teacher is an active facilitator, setting up educational contexts, carefully observing young children as they engage in learning, interacting when and where this encourages theory building, modifying the environment and activities based on what children do and how they interact, and supporting children in their interactions with others so that they learn how to resolve conflicts, express themselves, and interact constructively.

This reconstructualized role is far broader than that of the teacher as a transmitter of knowledge. Although the emphasis in the classroom is on what the *children* are doing and learning, the teacher is responsible for much of the orchestration that supports and stimulates children's learning. For example, in the second-grade classroom where the children built the "house," the teacher noted that as the children read their books, they showed intense interest in how the house was made. She then obtained the necessary resources, set up the activity, and facilitated the children's construction of the house. As they worked, she observed them and helped by providing needed materials or information. She facilitated their problem solving and encouraged them to expand upon their original plans. As this example shows, the teacher—while not the focus of the classroom and certainly not the focus of the children's attention—can play a critically important and complex role in facilitating children's theory building.

Our Conception of the Nature of the Curriculum

The curriculum for young children must facilitate children's theory building. Because of the way they learn, it cannot be prescribed and rigid but must be responsive to the needs, interests, and capabilities of the particular children being taught as well as the characteristics and dynamics of the children as a group. This is not to say that the curriculum should not be carefully planned—it must be even more carefully planned than usual with constant anticipation of the numerous ways in which children might choose to interact and work with the materials and activities presented to them.

Several issues in curriculum development force a reconceptualization of what curriculum entails. The first is the need for curricular integration. Subject-matter areas—such as reading, mathematics, science, writing, music—cannot be considered as separate units, either in terms of the time children spend on them or the activities as they are designed. Let us return once more to the house-building activity. Is this science? Is it social studies? Is it mathematics? In some respects it is all these things, because it offers children the chance to deal with each of these subject-matter areas if they so choose. In other respects it may be none of these. The curriculum is driven by the children; ultimately, it is what children *do* that determines the focus, and what children do in this activity is hard to classify.

The second issue is the evolution of the curriculum, which relates to the flexibility and lack of prescription of what is nonetheless carefully planned. An ideal curriculum is emergent and is responsive to what children do and what the observant teacher feels will stimulate theory building. Planning provides some structure and guidelines, but it is never rigid, for the best learning experiences are often those that are unpredictable and spontaneous. As we saw in the house-building activity, the diverse and integrated experiences that can occur in this process would never have arisen if the teacher adhered to a prescribed and rigid lesson plan.

WHAT THIS MEANS: RETHINKING WHAT WE DO IN CLASSROOMS

As we have suggested above, much of what the constructivist perspective implies is contrary to what we see in many educational settings today. The truly "nonconstructivist" classroom would be one in which there were few opportunities for children to interact with one another; where the teacher made decisions about what to teach on the basis of some external, specified, and prescribed criterion; and where the predominant mode of instruction was the teacher leading the group, usually talking but also presenting materials such as ditto sheets or "activities" that offered no choices for the children. An exaggerated picture, perhaps. Fortunately, educational settings are more varied than that, and the goal of this book is to offer you, as teachers and prospective teachers, a vision of the constructivist classroom. If you are currently teaching, such a vision can give you ideas about how to move toward being increasingly constructivist. If you are reading this book, you probably do not favor a highly "nonconstructivist" approach; you may not use ditto sheets, for example. On the other hand, you may still not know how to defend and expand those aspects of your classroom that are consistent with constructivism. Educational settings are under considerable pressure to be nonconstructivist, particularly (but not only) in some public school districts. In this book, we will be trying to give you reasons for the constructivist things that you do as well as ideas about how to do *more* that is consistent with constructivism. Even if you just move a little bit, you will be having an important impact on the children you teach, and perhaps even on the whole school—fellow teachers, administrators, and parents as well. If you are not yet teaching, we hope to give you some practical ideas for implementing a constructivist curriculum as well as the knowledge that your approach is both supported and shared by others.

It is, nonetheless, important to realize the profound implications of the constructivist perspective. For example, the true constructivist classroom would have some elements—such as increased staff, changed physical environments, and lack of formal, standardized testing—that are difficult to envision in many of our current educational settings. We will be discussing these implications in greater depth in Chapter 3. It is also important to create and keep in mind a vision of what the true constructivist classroom might look like, even if you feel frustrated at the profound changes it implies. Only through such vision can we work to improve the settings in which our young children are educated.

Constructivism and its implications regarding our conception of the learner, educational settings, the role of the teacher, and the nature of the curriculum is fundamental to our view of early childhood science education.

OUR DEFINITION OF SCIENCE

This book is about science education, and this means different things to different people. For some, science is associated with certain materials and content: magnets and magnetism, test tubes and chemical solutions, preserved specimens and microscopes. To us, these are but the surface manifestations of science. On a deeper level, the scientist is involved in the process of inquiry—of raising and trying to answer questions about the world in which we live. In fact, anyone who is studying anything in a methodical way is a scientist.

Science as defined by the surface manifestations that have come to be associated with it is often developmentally inappropriate and nonconstructivist in relation to young children. Although young children may be intrigued by certain materials such as magnets or mechanisms like microscopes, the extent to which they can understand the concepts involved in using those mechanisms and materials is very limited. As in other areas of the curriculum, educators have often made the mistake of taking what is appropriate content for older children and "watering it down" for younger ones. Our approach, instead, is to embrace those characteristics of science and the scientific method that are developmentally appropriate and to use the processes of inquiry as the starting point for early childhood science education. Science as we are defining it is thus quite broad; it

Young children are natural scientists, working very hard to build theories about their world.

involves experimentation, creativity, and problem solving, all of which come into play as children try to understand the world.

CONSTRUCTIVISM AND SCIENCE EDUCATION

This broad definition of science allows one to see science education as underlying every traditional curriculum area. Children are theory builders, and they do not build theories only when we decide that it is time to do so. Children are theory builders as they construct an understanding of print, figure out the meaning of numbers and how they relate to objects, try different colors and forms on the easel, interact with their friends, and figure out the meaning of reciprocity in friendship. Science in this very broad sense underlies the construction of all knowledge.

But we have an interest in focusing on science as it is traditionally considered as well. Today, children often grow into adults who have little or no grasp of scientific issues and little or no appreciation of or interest in science. At the same time, the field of science is, as perhaps it has always been, incredibly exciting and stimulating. Moreover, the potential applications of scientific understanding include some that may benefit humanity and others that can destroy the world. It is a particularly important time for us to consider the education of our children as contributing to the development of an aware, critical, interested, and sensitive "scientifically literate" population, with some subgroup of our children becoming the practicing scientists of tomorrow.

Thus we face quite a task. A constructivist approach to science education in early childhood would focus first and foremost on the ways in which young children think as they interact with the physical and natural world. Everything that we offer children with which to interact in the classroom must take into account their activities, materials, and environments in the broadest sense. Conversely, children's interactions are determined at least in part by the nature of their surroundings. Thus, we have to consider very carefully the nature of the knowledge that will be constructed in order to know how to present activities, select and introduce materials, and set up our educational environments.

In subsequent chapters we shall take three knowledge bases—chemistry, physics, and biology—and consider carefully what they mean, what young children can know, and how we can best provide children with opportunities to experiment in each of these areas.

OTHER APPROACHES TO SCIENCE EDUCATION

There are several different ways of looking at science education. Some of these overlap; that is, the same programs or people could subscribe to more than one of the following beliefs or practices. We shall present some of the beliefs that have come and gone in the field of science education and discuss how they are similar to or different from the perspective presented in this book.

A Common Misconception: That "Real" Science Education Depends on Advanced Abilities

Some science educators believe that science requires formal, abstract thinking; that is, that children must be able to think in terms of abstract "cases" and "variables." They argue that science by its very nature is inaccessible to young children; therefore we shouldn't spend too much time on it. Exposure to some of the content of science, which they will study in greater depth later on, is sufficient.

This belief probably accounts at least in part for the fact that many students in the primary grades receive little or no science education in the public schools. You can also see that if you subscribed to this belief, you would take the existing science textbooks written for older elementary students and just water them down for the younger ones. This is precisely what is done by many textbook companies.

Quite clearly, we do not agree with this; otherwise we would not be writing a book that focuses on science for young children. The issue is what is meant by "real" science. "Science" as most people, including some scientists, conceive of it includes a large body of factual knowledge and some methods that are used to obtain, verify, and extend such knowledge. It is true that some of this knowledge, some of the ways in which it is organized, and some scientific procedures are inaccessible to the mind of the young child, who cannot yet deal with the required level of abstraction. While we readily grant this, we also contend that this aspect of science is only *part* of what is in true scientific inquiry. Real science incorporates many things to which young children are most particularly open: creative thinking and problem solving, experimentation and invention. Many of the processes that the practicing scientist uses are seen, in microcosm, in the way young children construct knowledge in *all* domains, as we have argued, not just in what we term science. You could, in fact, argue that what young children do is real science, and what bored adolescents do in dry science classes is but a weak reflection of this.

Another Common Misconception: That a "Hands-on" Curriculum is Enough

This belief is grounded in the idea that children "learn by doing"—by manipulating and using real objects and by actively participating in the processes of learning.

Unfortunately, there is a broad range of interpretation as to what this may mean for curriculum. For some educators, a hands-on curriculum is one that enables children to be actively involved in constructing knowledge—one that is driven by *their* questions that gives them considerable access to materials and many choices as to their use. These interpretations are all consistent with the ideas to be presented in this book. Others define "hands-on" as meaning, quite simply, touching or manipulating objects as a way of learning about them. For example, children being taught about skeletons may be given the opportunity to touch, hold, and play with real bones. The key here is that the touching of the bones is a means to the end of teaching children about them. In fact, the act of touching the bones may only serve to reinforce the "lessons" taught to the children, usually by the adult, regarding bones. If you stop to think about it, what can you actually learn about bones simply by touching and manipulating them? In a constructivist setting, children may be given the opportunity to play with and experiment with bones. From such play may arise some spontaneous curiosity among the children, as well as questions that might generate other activities. The hands-on nature of activities is an important part of a constructivist curriculum, but in and of itself this does not make an activity a constructivist one. This is because the constructivist sees the essential activity as what goes on in the child's head, not in his or her hands. With young children, physical activity and manipulation is often a necessary part of mental activity, but not always. Think of the young child—Billy, for example, in the scenario at the beginning of this chapter—intently watching other children involved in their own experimentation, who suddenly has his own wonderful idea and acts on it. Children need to be active, yes, and they need opportunities to manipulate and experiment with real objects. But this in itself is not the definition of a good activity.

"Process" Versus "Content" as a Focus of Science Education

We might divide the field of science education into two camps—those who focus on process and those who emphasize the content of science. The processes of science involve inquiry, experimentation, systematic description, and hypothesis testing. The content of science, according to those who make such distinctions, is the information that is specific to a given domain of knowledge. Take the field of botany as an example. Botanists use various tools that are common to all scientists—methods for exploring, describing, and developing knowledge in the field of botany. These tools include experimentation, categorization, and observation. All these are tools that are also useful to the geologist, the physicist, and the medical researcher. These tools are the *processes* by which the botanist comes to know things. But the botanist also has knowledge particular to botany, what we would consider the *content* of botany—names of plants, descriptions of processes of plant reproduction, and particular techniques that are used primarily by botanists in their work.

For the constructivist, the distinction is an artificial one, because children's construction of knowledge involves both content *and* process. Moreover, the two are very closely connected.

Particularly for younger children, a focus on process in the primary grades would emphasize methods and tools that cut across content areas. That is, content would be essentially secondary, because learning to observe *anything* or the systematic description of *anything* could be a curricular focus. Focusing on process does not imply that content is not important; indeed, you are incorporating content when you focus on process. Rather, content is not the primary focus.

Consistent with a process approach one might, then, recommend that children be encouraged to make and share collections as a way of learning skills of categorization and description that, presumably, could generalize to any field of later study. Similarly, children might be encouraged to make predictions about events that will occur. Before going on a field trip, the group would generate predictions about what they might see and do on the field trip. The process of predicting is the point of the activity. Yet in the process of predicting and categorizing, children would be dealing with a content area of some sort.

A focus on content would emphasize the specific knowledge of a content area. Learning facts in an interest area might be the focus of the child's activity. *Labeling* would be of interest to the educator who emphasizes content. Such an educator would stress the names of things, the terms used to describe processes, and particular kinds of procedures. Continuing with the example of botany, the teacher would use the names of trees and plants, books would identify and classify plants by species, and activities would focus on the learning of those labels.

To better understand how content and process are intertwined, let's consider what the constructivist teacher might have done with a group of first-graders to capitalize on the following event.

■ Jim, who just stepped on a decomposed log, says, laughing, "Hey, my foot sunk down in this log!" "Jim's foot sank in this log!" repeats the teacher, so that the whole group can hear. Some of the children quickly gather around to see. The teacher squats down to inspect the hole made by Jim's foot. Picking up some of the crumbled log in her hand, she asks, "Any suggestions about what changed the log?" This question generates quite a few varied ideas from the children, ranging from "The bugs chewed the log up" to "The rain keeps melting things on the ground." The teacher suggests that they inspect the log and the ground around it further to see if they might notice some more "decomposition" in the wood and soil. Jim says, "I know! Let's use the microscope." The children gather up samples to take back to the classroom.

Upon returning to the classroom and using the microscope, the children discover several different kinds of bugs and begin looking through some "bug identification" books that are on the bookshelf.

As you can see, the teacher helped to focus the children's thinking on change and facilitated their thinking about causality. In so doing, she also dealt with some age-appropriate content, such as vocabulary related to decomposition.

A well-thought-out constructivist approach to early childhood science can blend an emphasis on process with appropriate science content to encourage children's experimentation and theory building.

Chapter
2

A Constructivist Curriculum Model for Science

■ In this first grade classroom there is lots of activity and children are involved in various learning centers. At one table there are numerous wheels of different sizes and shapes made of play dough that the children made yesterday. One of them, Sarah, checks on the wheels she made, confirming that they have hardened. She takes her wheels and sits down at a table with a selection of straws and cardboard cartons of different sizes. After sorting through the cartons, she selects a small one and announces, "I'm going to make a funny car!" Sarah pokes straws through the holes in the cartons and makes two axles, then selects two round wheels for the front and two square ones for the back. As she moves the car around the edge of the table, she puts her fingers lightly on the back square wheels as they slide along the table surface.

Sarah stops her car abruptly and begins to smile. She takes off the two rear wheels and, after searching quickly through her pile of wheels, selects two egg-shaped ones and puts them on the rear axles. Now she rolls her car along the table and then on the floor, laughing and calling to the other children as it bumps along.

This activity is science education from a constructivist perspective. Why? Because it is providing children with a developmentally appropriate opportunity to engage in experimentation with the physical world. Sarah is self-directed, bringing her own ideas about movement to an activity that lends itself to many different types of experimentation. This experimentation is done in a context that both allows for and encourages social interaction; it lets children explore their ideas individually or cooperatively. These elements of the activity—self-direction and choice, social interaction, and active manipulation of materials—are congruent with the developmental needs and capabilities of the young child.

The constructivist perspective described in Chapter 1 implies rethinking the way we develop and implement curriculum. In this chapter, we will present a curriculum model that can help you to design activities that will generate the kind of experimentation illustrated by the above examples. After presenting the curriculum model, we shall discuss some of the issues raised by it—issues such as the role of the teacher in this child-directed context and the role of conflict and contradiction as children wrestle with the questions posed by the materials and activities. Finally, we shall discuss how the curriculum model reflects our goals for children.

THE CURRICULUM MODEL

A curriculum model is a framework that enables the teacher to make decisions about what will go on in the classroom. It can help the teacher to choose materials, select and evaluate activities, and coordinate many different classroom experiences.

As we discussed in Chapter 1, a constructivist perspective focuses our attention on the *child's* contribution to the construction of knowledge. Constructivism is based on the idea that children are actively engaged—naturally and without the aid of direct instruction—in building theories about the world and the way it works. From a constructivist perspective, children are natural scientists, and, given the opportunity, will engage on their own in experimentation and problem solving. The role of the teacher is to provide contexts within which such experimentation can occur and to facilitate theory building by providing helpful materials and experiences. The processes whereby children acquire and extend their understanding are of critical importance.

It follows, then, that a constructivist curriculum model should be derived from the child's own thought processes rather than some content-oriented topic or theme arbitrarily chosen by the adult. And since children are viewed as actively inquiring natural scientists, our curriculum model should encourage experimentation. Materials and activities must allow for many possibilities. The curriculum model must provide guidance by creating learning situations that allow and encourage diversity. Such settings will let children produce and test many different ideas and hypotheses, creating an environment that actively supports theory building.

Let's think more specifically about what this process of theory building looks like when children are engaged in an activity. Consider, for example, an activity that gives children the opportunity to combine materials. Children come to a project table at which there are containers with different substances, such as flour, water, sand, salt, and oil. There are also empty cups, spoons, basters, and medicine droppers. There are many ways in which children can combine these substances and many ideas that they can explore. A child can, for example, be interested in adding different amounts of water to flour, comparing the resulting differences in consistency. Another child could be engaged in exploring how flour combines differently with different liquids. Let's look at this activity as it is implemented in a preschool classroom.

■ Lucas, age 4, sits down at the table and looks at the containers of flour, salt, sand, water, and oil. He dips his finger into the flour, holds it up in the air, and blows on it. He laughs as the flour is blown off his finger and then dips his finger into the salt. He looks at it, noticing that nothing has stuck to it. Lucas reaches for the pitcher of water and pours some into an empty bowl. He sloshes his hands around in the water for a few minutes and then scoops a few spoonfuls of flour into the water, stirring it quickly around with a spoon. Focusing very intently on his efforts, he lifts up spoonfuls of the mixture and, tipping the spoon, lets it drop back into the bowl. He repeats this a few times and then puts one hand under the dropping mixture and lets it fall into his hand. Laughing, he rubs his hands together with the mixture, holds them up in the air for another child to see and says, "See my goop!" He then plunges his hands back into the liquid in the bowl.

Lucas adds more flour and then more water. He then repeats the pouring/feeling process. After working like this for a while, he dips his hands into a large bowl of plain water and swishes them around to clean them. He then dips his finger into the salt and inspects the salt-covered finger. He blows on the finger and notes that his breath does not blow the salt off. Then he dips his finger in water. Finally, Lucas takes another bowl and pours water into it. He begins scooping salt into the water, feeling the mixture with his hands after each addition.

Lucas is actively experimenting in a variety of ways with the properties of the different substances and combinations. If you look closely at his actions, you can see that he is asking several questions as he explores the materials. For example, as he puts his finger in various substances, he notices that some substances stick to his finger and some do not. Adding the flour to the water, he explores the consistency of the "goop" and then experiments with what happens when he adds more flour and more water to the goop. Woven into this inquiry is the continuing question of what will stick to his finger, since he notices that when his finger is wet, things stick differently.

Lucas is posing his own questions, and his actions lead him quite fluidly from one experiment to another. Similarly, in the example at the beginning of the chapter, children are involved in constructing cars with different combinations of wheels and then altering them. Sarah plays with and compares the cars she constructs, experimenting with what happens when the wheels are of different shapes and asking how this affects the way the car moves. The combination of delight and seriousness of purpose that we see in both Sarah's and Lucas's experimentation is facilitated by materials such as these, materials that suggest experimentation and provide variety in a self-directed context. In both examples, children have many options for experimentation and can choose from a range of more or less complex concepts to experiment with, testing out many different ideas.

Now we have an initial idea of the overall goals of a constructivist curriculum model, and we are beginning to see what the resulting activities should look like. Our task is to provide a supportive environment in which children can ask their own questions and have the means to look for answers. The curriculum model itself reflects what we expect children to do when they engage in activities—it is based on *questions*.

These are not questions that teachers ask but questions that children ask as they engage in activities. If you try to look at the examples of activities we have given from the *child's* perspective, you begin to see the different questions that the child is asking as he or she engages in experimentation. Some general categories of experiences begin to emerge, categories based on the particular kind of experimentation that the child is asking as he or she engages in experimentation. We can put these categories of experience into the form of the question that the child may be asking. Going back to the example given earlier, of children engaged in constructing cars with different types of wheels and experimenting with the shapes and sizes of the wheels, the general question is "How can I make it move?" And where the children are engaged in combining different substances, the general question is "How can I make it change?" Such broad questions—"How can I make it move?" and "How can I make it change?"—can serve as organizers in curriculum planning. The activities that can be developed around these questions will be linked on the basis of the types of experimentation and theory building that the activities and materials will encourage.

In speaking of questions as organizers of the curriculum, we are not referring to *verbal* questions asked by adults or children. Young children can demonstrate a great deal of understanding and inquiry without ever saying a word! When Lucas was trying out different substances and exploring their various properties, he did not verbally ask the question "What is happening here? Is the salt sticking to my finger and the flour blowing off it?" Yet his actions, his repetitions and variations, and his intensity of interest are all indicators of active inquiry. Children do much of their most important work and play without saying anything about what they are learning. Nor does a curriculum model organized around questions imply that teachers are asking questions of children. Quite the contrary! Teachers are creating activities and an environment that encourages inquiry—experimentation and theory building. The teacher's role, after he or she has set up these opportunities, is to be an active observer and facilitator of inquiry but not to interject questions unless they are appropriate. We shall have more to say about the teacher's role later and will devote Chapter 4 to the techniques that teachers can use to decide whether and how to be involved. For now, remember that when we pose questions, we do not mean them to be actually asked by teachers or necessarily verbalized by children.

QUESTIONS TO GENERATE EXPERIMENTATION

In this book, we'll be describing some general questions on which to base an early childhood science education curriculum from a constructivist perspective. The three questions that we shall focus on are:

How can I make it move?

How can I make it change?

How does it fit or how do I fit?

Each of these organizing questions corresponds to a primary category of scientific inquiry. "How can I make it move" involves experimentation that explores some basic concepts in the area of physics; "How can I make

"How can we make it move?" reflects these children's experimentation with inclines and rolling objects.

it change?" explores concepts in the area of chemistry; and "How do I fit or how does it fit?" explores concepts in the area of biology.

"How can I make it move?"—our organizing question in the area of physics—incorporates a wide range of curriculum activities and materials that let children experiment with the movement of objects. Thinking about the different ways that children could ask the question leads us to design different materials and activities. Thus, for example, the child could ask the question "How can I move it by tilting?" The child may, for example, construct and play with mazes that must be tilted in order to move a ball. This could provide an opportunity for experimentation with the underlying question. As another example, the question of "How can I move it by rolling?" could be stimulated by providing an opportunity for children to experiment with balls and inclines. In Chapter 5, we shall describe the curriculum that can be generated by using this question as an organizer.

How can I make it change?—the organizing question in the area of chemistry—underlies many of the young child's experiments. Transformation of materials characterizes much of the young child's spontaneous play. Children's interest in and active involvement with such materials as play dough and water, as well as their interest in construction toys, reflects the

"How can we make it change" is the prevailing question of these children's experimentation with tissue paper, glue, and water.

pervasiveness of this question. By designing materials and activities that encourage experimentation with transformation, we can facilitate children's theory building. Chapter 6 will describe some approaches to this curriculum area.

"How do I fit or how does it fit?"—the organizing questions in the area of biology—explore what we call ecological perspective taking. How does the child perceive himself or herself in relation to the rest of the natural world? How do my actions affect the world around me? For example, how does the rabbit respond when I sit still and hold her in my lap: is she calm and relaxed and not trying to hop away? Certain types of experimentation in this domain are often not possible or desirable; for example, we would not want children to see what happens if they screamed at the rabbit. In this domain, perspective taking is emphasized; this involves reflection on the possible effects of your actions on the world around you. Such reflection is facilitated by the development of sensitivity, empathy, and appreciation for the natural world, including an awareness of transformations that occur there. This curriculum area will be described in more detail in Chapter 7.

Being both attentive and gentle with the insect she has discovered is part of the curricular focus "How does it fit?"

HOW THIS CURRICULUM MODEL IS DIFFERENT

The curriculum model presented above has a different focus than do some traditional approaches to curriculum development. There, curriculum in early childhood programs and early primary science education may be organized around content-oriented themes, or topics. Some content-oriented themes that might be familiar to those who work in preprimary programs are "transportation," "fall," and "teddy bears." Likewise, in more traditional approaches to science education in the public schools and some preschools, familiar content-oriented themes might be "matter" and "the weather." Such themes are derived from the "content" of the activities that are planned for children.

As an example of a traditional theme approach, take the theme "dinosaurs" for a group of 4- to 5-year-old children. Activities planned for one or two weeks would be designed around that topic. For example, pictures of dinosaurs would be displayed around the room. Toy dinosaurs would be placed in the block area to encourage symbolic play. At the art table, dinosaur stencils might be made available. Books with dinosaurs in them might be selected for the library. Games like lotto and puzzles that have dinosaur pictures might be put out in a game area.

To see how these activities can be linked only by subject matter, try substituting the word "vehicles" for "dinosaurs," and you have the theme "transportation." There is really no connection between the activities other than the particular content. From the child's perspective, the activities are each very different—not connected in any way except that there are particular images and labels associated with them.

This is not to say that the individual activities that teachers might choose to go with content-oriented themes might not be enjoyable, developmentally appropriate, and even exciting. It *is* to say, however, that as a curriculum model, the use of content-oriented themes may not promote integration and won't focus the teacher's attention on the child's perspective. Instead, content-oriented themes make the teacher focus on the subject matter, which often reflects social arbitrary knowledge.

There are several important contrasts, then, between content-oriented themes and our question-focused type of curriculum development. First, in *planning* a curriculum around the questions we have suggested, the teacher's attention is directed to the child's thought processes. It becomes important for the teacher to try to put himself or herself in the child's place so as to understand how the child will approach the materials and experiment with a particular activity. Thinking about the experimentation that children can engage in helps us, as curriculum planners, to anticipate some of the many possibilities of experimentation and to encourage the testing of hypotheses. In this way Sarah's teacher anticipated that Sarah might be interested in experimenting with the size of the "car" she was building. In order to make such experimentation possible, the teacher had different sizes of milk cartons available. In contrast, the content-oriented themes a teacher might choose do not necessarily encourage the teacher to see things from the child's perspective. Rather, the teacher turns to the topic itself for ideas: What do dinosaurs look like? What games use dinosaurs in them? Which books have pictures of dinosaurs?

Second, *integration* of the curriculum is achieved when one uses the curricular questions we have proposed—integration from the important perspective of the child. Presumably, the reason we have a curriculum model at all is so that there will be some coherence in what we do with children across time and across activities. In this case, the child's own activities and inquiry provide the continuity—not the adult-selected subject matter. An activity related to the one involving car construction would be to make available objects of different shapes and various inclines. Then, the children's experimentation would provide continuity with the other activity.

In contrast, an adult-selected, topically related activity might focus on the topic "cars" and might provide a collage-making activity, with pictures of cars available. The collage-making activity would be related *only* by topic and not by the child's actions.

The question-focused approach to curriculum development is different in that it takes as its starting point the questions children ask. However, it is not necessarily inconsistent with the views of those who have proposed the "project" approach to curriculum development, namely Katz and Chard (1989) and Gamberg and co-workers (1988). Nor is it inconsistent with the approaches to integrated curriculum taken by Pappas, Kiefer, and Levstik (1990). But it does have a slightly different starting point in order to explicitly develop integrated activities around the three traditional content areas of science. And, whatever route you take to create a constructivist curriculum, you need to be wary of a disproportionate focus on social arbitrary knowledge.

WHAT SORTS OF QUESTIONS ARE CHILDREN ASKING?

To look more closely at the continuity that this curriculum model brings to children's experiences, let's consider the kinds of questions children ask as they engage in such a curriculum.

First, note that the questions derive from the children's own experiences and observations: "What will happen if I roll this ball down the incline?" "What will happen when I mix tempera colors into the play dough?" Because children experiment actively with the world and are keenly interested in observing it, they are open to all the events and occurrences around them, particularly when their actions are directly related to those events. This "sense of wonder," to use Rachel Carson's phrase (1987), is not just an openness to sensory experience but involves the active processing of information.

It is paralleled by a continual puzzling about what is experienced. Children are rarely satisfied with answers to their questions because their experience only serves to generate new ones. The child, just like the scientist who is never satisfied with an answer, continually searches for new knowledge and poses new questions. Each new understanding brings new things to learn, opening up new areas for exploration and experimentation.

These two characteristics of young children—wonder and puzzlement—make theory building a process of continual ups and downs, successes and failures.

This experience of theory building and experimentation is delightfully rich and diverse. The teacher must be observant, insightful, and flexible in order to provide a stimulating and supportive environment. The remainder of this book is devoted to how these goals might be achieved, covering both the "why," or theory, that underlies constructivist education and the tools the teacher will need to implement the theory in the classroom.

PART
TWO

Setting the Stage

Chapter
3

Creating a Constructivist Learning Environment

Picture two classrooms. One is loosely divided into areas by 2-foot partitions, low shelves, rugs, and tables. A quick look around the room reveals a variety of learning centers with different physical features. For example, in one corner is an area with low bookshelves filled with books, large pillows, a small low table, and writing implements. In another area is a water/sand table near shelves with bins of containers, tubes, sieves, and scoops. Half of this classroom has been left open—with blocks, ramps, and arches as well as bins of toy animals, vehicles, and balls. There are sign-making materials around the perimeter. The room also includes a table to accommodate table-top activities. Hooks and a pulley system are attached to the two-by-fours that run along the wall and ceiling at 6-foot intervals.

Three-fourths of the other classroom is filled with rows of tables and chairs. Shelving containing books, writing materials, games, and art materials runs along the perimeter. In the front, next to the teacher's desk and facing the chalkboard, is a small carpeted area. At the rear of the room, tables have been pushed aside to make a small open area for a marble rollway. Bags of cut tubes, boxes of blocks, marbles in a container, and tape are set against the wall.

Two very different classrooms! A learning environment for constructivist early childhood science experiences can be designed in many ways and to different degrees. In this chapter we shall discuss ways of creating learning environments that can help you do what *you* want to do, and enable you to build some of the ideas in this book into your own teaching. We shall describe aspects of the physical environment to be considered—including the use of physical space and furniture—and the materials and objects that are to be made available. Considerations related to the social context of learning will also be described. This discussion will include not only ways of facilitating self-direction—both through the accessibility of materials and through the socialization of children for self-direction—but also methods of encouraging children to interact with each other. Finally, we will consider some of the constraints and challenges that teachers may be faced with in attempting to implement the ideas presented in this book. Some ideas for handling such constraints and challenges will be presented.

THE PHYSICAL ENVIRONMENT

Overall Physical Space

There are a couple of principles that guide us in considering the overall physical environment. First, though, we believe that good constructivist activities can take place in almost any environment with appropriate provisions. Teachers who want to facilitate theory building and children who want to test theories are very adaptable when it comes to the use of physical space. That said, however, you may also encounter a lot of stumbling blocks to carrying out your ideas. These can often be remedied by *where* you are working and playing. A carefully thought out physical environment that embodies the principles described in this section can facilitate the smooth flow of play and activities. Thought given to the use of physical space can help solve numerous behavioral problems that teachers often face. Therefore it is essential to consider how the overall physical space can contribute to our constructivist goals.

Principle 1: Allow for good traffic flow and easy, unmonitored movement from area to area.

One of the keys to a constructivist classroom is that children have freedom of choice and the ability to go from area to area without asking to do so. In the ideal classroom, as we will describe later, the physical space allows children to incorporate elements from one area into another; in planning the rooms, therefore, we must consider how that flow can occur. Always think of it from the child's perspective too—sometimes a pathway that is clear to a "tall" adult is not at all visible to a 3-foot child.

Also, as is true with any classroom, it is important that traffic flow not disrupt ongoing activities; for example, a clear wide path should not infringe on the block area. This is again particularly important in a construc-

Smooth traffic flow and accessibility to materials support children's scientific experimentation.

tivist curriculum. Why? Because when traffic flow disrupts activity, conflict occurs that must often be monitored by an adult. Since the goals are to have children deal with problems themselves, we want to minimize unnecessary conflict.

Principle 2: Allow for as much flexibility in the use of physical space as is possible.

Whatever you have, let it be flexible, adaptable to a number of different uses, movable, and transformable. Because this constructivist curriculum will be responsive to what children are doing, it is difficult to predict exactly what you will need at any given time. So what you have must be changeable to respond to your changing needs and those of the children.

For example, you might have your dramatic play area partly separated from the rest of the room by a room divider. This provides flexibility in the use of that space because the room divider is movable. Thus the dramatic play area can become larger or smaller, as necessary, to accommodate different types of play at different times. At one point the dramatic play area might be a transportation area, requiring quite large spaces for gas stations, repair areas, and drive-in banking. At another point, when the dramatic play area might be a restaurant, it could be smaller and cozier.

Flexibility is enhanced when walls are minimal and room dividers are movable. Flexibility is enhanced when parts of the room can be moved around—when the easels for art aren't fastened to the wall, so that they always have to be in that place. Flexibility is enhanced when floor coverings are versatile or movable. For example, large slabs of linoleum that are taped down over industrial carpeting can be picked up and moved elsewhere if necessary.

Remember, these principles are goals—you may not be able to *meet* them given the limitations of your environment. But even the most limiting rooms have *some* floor space that can be made available, and it is amazing how easily and happily all children, even second-graders who are accustomed to sitting at desks all day, can work on the floor. Look around your room for any area that you can adapt for constructivist activities, and try to be flexible yourself!

Furniture

The principle of flexibility is also important when it comes to furniture. Classrooms that have individual desks with chairs offer few options for their use. Even worse, and still found in some classrooms, are desks or tables that are bolted to the floor! Less extreme are classrooms that have large tables, particularly very large ones, that do not easily lend themselves to multiple uses.

An example is the type of large, round or rectangle table often found in preschool classrooms. This table, seating maybe six children, is usually seen with an activity on it, with six individual "portions" or setups on it: a glob of play dough with some tools at each place or puzzles at each place. The table itself is really too wide to allow for a great deal of interaction across it. Thus, a large table lends itself to individual activity, often parallel play, with interaction between children next to each other. Large-scale co-operation or combined activity is rarely seen at such a table.

Smaller tables, those that seat no more than four children, or even *no* tables, may be much better for the types of activities described in this book. In cases where a solid surface or delineated area is needed, pieces of masonite that can be stacked or put away can be used.

Versatility is key in determining furniture choices. For example, instead of a wood toy stove and refrigerator for the dramatic play area, you could choose a couple of large wooden boxes that could serve as stove and refrigerator but could also be used for any number of other purposes in imaginative play. Little benches instead of chairs might also lend themselves to other uses (serving as shelving or small-scale tables). Modular units that can be configured in different ways would be particularly valued in a constructivist classroom.

The teacher in this classroom transformed the learning environment quickly and without extra equipment by turning a table upside down.

OBJECTS AND MATERIALS

A constructivist classroom does not have to be filled with expensive materials and toys. Some of the very best activities are those developed from found materials—a long clear plastic mailing tube, a set of cardboard boxes, a box of spools, large odd-shaped sponges used for packing dishes—these can serve as inspiration for some of the best activities. In this section, we would like to share our experience with some of those noncommercial objects and materials that come in particularly handy as you develop and implement activities. (Note: In the appendix we give a list of materials that also includes those recommended in subsequent chapters.)

Good Things to Have Around

Clear Hard Plastic Clear hard plastic is an excellent material because it is durable and makes processes visible to children. Clear plastic tubes, for example, allow the children to observe the varying movements of the ball as they alter the incline of the tube. A framed, large piece of clear hard plastic (approximately 2 feet by 4 feet) may be used either vertically or horizontally for processes such as finger painting, which may be viewed from either side to encourage perspective taking. Small pieces may be used for the same purpose but held by individual children.

Tubes and Tubing   These may be collected from a variety of sources (parents are a great help with these items) and are useful for both planned activities and as "extenders" to blocks. All sizes are good, from toilet-paper tubes to giant paper rollers. A tube may be cut in half vertically to make the movement within it visible, or it may be left whole. Tubing of all widths is also useful and provides the added feature of being bendable if it is thin enough.

Containers   Collect containers in a variety of sizes with multiples of the same size. Again, because we want children to be able to observe the interactions between physical properties of the materials and their actions, clear containers are more desirable than solid ones.

Boxes   Boxes may be used for small- and large-scale construction by the children. They are light and readily available. Large boxes may be used intact, altered slightly, or cut apart for large pieces of cardboard.

Sand and Water   These are some of the most exciting and flexible materials for young children. It is good to provide opportunities to use them together in varying proportions as well as separately.

Play Dough   Play dough ranks right at the top of the list of good constructivist materials. It is a solid material, yet completely flexible. It may be used alone or supplemented with implements. Avoid implements that create predetermined representations, such as cookie cutters. Instead, select implements that are open-ended, focus children on the process of transformation, and allow for creativity. Examples would include rolling pins, dull knives, empty thread spools, strings of beads, and scissors.

Blocks   A complete set of wooden unit blocks with planks and arches is durable and multidimensional. Smaller blocks may be useful to extend block play to a different scale as well as to be substituted in environments where large constructions are not practical.

Pulleys and Hanging Things   The ideal classroom would have reinforcements around the room where pulleys and hooks could be attached. These open up possibilities for developing activities of many kinds that might involve temporary constructions, enabling the teacher to hang or suspend things from the walls or ceilings. However, even one pulley securely placed in a convenient spot may be useful for multiple activities. Make sure it is working easily. For example, if the pulley is to be easily used by children, the rope running through it must be of a diameter appropriate to the pulley's size.

How to Make Things Accessible

In a constructivist classroom where children's self-direction is valued, you will want to minimize their reliance on adults to get the things they need. As you plan activities, you must think carefully about accessibility. You will want children to have free and constant access to some materials and objects. Other materials and objects will be made available only for particular activities or at particular times.

In your setup for activities, you must make sure that everything children need is readily available. Low shelving on which materials can be placed or benchlike tables can make that possible. The flip side of accessibility is that you *don't* want things to be accessible that you don't want the children to use in a particular activity. If you don't want them to mix tempera paint with shaving cream, then don't leave the paint readily available near the shaving cream activity. Accessibility encourages children to use the available materials. Flexibility is still the key, though, because you don't want to stifle interesting ideas that you may not have foreseen. If you don't want to encourage the mixing of paint and shaving cream, you won't put them together. But be flexible enough to ask yourself "Why not?" Therefore, if the children ask for the paint and come up with the idea on their own, you may decide to respond by making it accessible. It *is* a logical extension of the activity; maybe you were even planning on putting it out in a few days.

Always think of how the materials and objects appear to the child. Sometimes it is helpful to think about how overwhelming it can be to have too much available. A reasonable number of choices, clearly accessible, is better than an overwhelming array. For example, in a game area, you would not want to have all your games always available. Instead, you would choose a few to have out and rotate two or three of them a week for variety, keeping one out longer if it seemed appropriate.

Rotation The rotation of materials and objects always stimulates new ideas. While certain things are always needed—pencils, markers, scissors, and blocks—many others can be routinely varied as part of your curriculum planning. A set of plastic tubing by the water table can be rotated with a set of containers. Sometimes the two sets can be combined. Similarly, a set of rollers to be used with the blocks can be rotated with other props, like toy dinosaurs, each encouraging different types of play.

Reciprocity   Reciprocity—which is defined as "mutual dependence, action, or influence"—in this context has to do with encouraging the use of objects and materials across areas. Reciprocity is not an overall principle; there are some reasonable restraints to consider, such as not allowing children to use the paint on the books. But it is a consideration that partly determines how you socialize your children to the classroom, what is permitted and why, and how you set up physical space and make materials available.

In general, we feel that reciprocity should be maximized where it is reasonable and, in fact, should be stimulated and encouraged in many ways. Why? Because reciprocity is one way of giving children more fuel for their self-directed problem solving.

Consider these two situations. The first follows a child in an environment where reciprocity is *not* allowed or encouraged.

■ Seven-year old Jennifer moves a marble with her hand along a long unit block. After several minutes of this, she tips up one end of the block, places the marble at the top and releases it. In doing so, she is attempting to keep the marble on the block as it rolls. Each time, however, the marble falls off the side of the block. Jennifer gets two other long blocks and tries to hold them together with the other block to create a trough for the marble. Unable to manage both holding them *and* retrieving the marbles, Jennifer stops her experimenting and goes to another "learning center."

Now picture Jennifer in an environment where reciprocity *is* fostered.

■ This time, after being unable to hold the "trough" and also roll the marble, Jennifer goes to the library area and brings back some pillows. She uses the pillows to support the long blocks so that the rollway trough is free-standing. However, as she turns to get the marbles, she accidentally bumps the pillows and the blocks separate, making cracks for the marbles to fall through. Bobby, who has been building a block tower nearby, pauses to watch. Noticing what just happened, he goes and gets some tape and hands it to Jennifer, saying, "You could tape it." Both Jennifer and Bobby tape the blocks together and make the trough secure. The children roll marbles down the trough for several minutes and then decide to expand their project and create troughs across the block area. As cleanup time approaches, Jennifer goes to the writing tables and brings back some cardboard, pens, and popsicle sticks. Together, Jennifer and Bobby make signs and secure them to the bocks. The signs read: "PLEZ DU NT MVE."

Because the second environment was adaptable to the children's needs during experimentation, Jennifer and Bobby were able to work together to follow through on their ideas and to expand their experimentation into other areas, including literacy.

THE SOCIAL CONTEXT

The environment for learning that is intended to facilitate the construction of knowledge must take into account the importance of the social context in which knowledge is constructed. As we discussed in the first chapter, social interaction facilitates children's theory building. In addition to exposing children to various ideas and perspectives, social interaction requires them to coordinate their own perspectives with those of others if the "play" is to proceed. Several aspects of the social context must be considered in setting up a learning environment.

What is the social context of learning? It is the complex web of social influences that affects what children do in the classroom as well as what they come away with. These social influences include direct social interaction—that is, children's interactions with other children and with adults. This also implies allowing children *not* to interact with others when they need to be alone—to solve their own problems or obtain resources without adult assistance. It means, in general, recognizing the child's right to be self-directed, or autonomous, as opposed to other-directed, or heteronomous.

Keep in mind that all these aspects of the social context of learning come together in what we might call the classroom "climate" or "culture" in which children work and play and that these aspects are complexly intertwined.

ENCOURAGING SOCIAL INTERACTION

Expect Reasonable Noise and Movement

The first step is to permit reasonable noise and movement in your classroom. For some teachers this is difficult. In some classrooms, children are

A narrow table with shared materials in the center encourages social interaction.

told that there are "inside" and "outside" voices. In some classrooms, talk that is not directed toward answering questions is automatically considered "off task."

Excited, involved children are often noisy, active children. When children interact with each other, they often make quite a bit of noise, sometimes talking loudly, calling across the room, interrupting each other, arguing, discussing intently. They often move from place to place quickly, running to get materials and resources. And young children in particular, if constrained in the way they are to conduct such interaction, have difficulty monitoring their own behavior.

So if interaction is to occur, expect noise. We said *reasonable* noise. Reasonable noise is noise that does not overly offend others and that is not mean or hurtful. It is reasonable to have constraints on noise that is offensive or that actually interferes with social interaction.

But also look carefully at how that interference occurs. Sometimes issues related to the design of the physical space lead to unreasonable noise or movement. For example, an area of the room designed for fairly rowdy dramatic play should not be directly next to the quiet book area, where children might be disturbed. You, as the teacher, might be able to arrange things so that noise that is unreasonable in one context is reasonable in another.

You can also redirect unreasonable noise or movement. If children are screeching as they play a game and it is filling the entire classroom with noise that effectively stops other activity, you can suggest that the play be done outside or in a separate room where it won't disturb others.

Look at the blend of activities that you are expecting to occur simultaneously. When children are themselves actively involved in an activity, whether it requires quiet concentration or not, they can screen out a great deal of interference. But don't expect a large group of children to engage in a boring task, such as doing a workbook exercise, while exciting noisy activity is going on in another part of the room. Sometimes the very fact that the children are distracted can be an indication to you that they do not find the available activities challenging, exciting, or even interesting!

Look at the Activities That You Have Designed or Made Available

A key to encouraging social interaction is to provide materials and activities the children will find difficult to deal with all alone or that just aren't as much fun that way. Blocks, which children can use alone or with others, seem to encourage social interaction. So do larger-scale construction toys, dramatic play, large objects such as pulleys and levers that require two or more children just to use them, and activities that involve transporting materials from one place to another.

Sometimes the lack of enough materials gently forces children to work together. For example, a limited number of wheels in a building set might "suggest" to children that they work together to construct a vehicle. *One* water pump in the water table might actually be better than several, necessitating cooperation and sharing in its use.

An entire activity can be designed to require or suggest interaction. Extending the water-play example you can provide *very* long tubing and ways of moving water from one tub to another. Long tubing is difficult for one child to manipulate without help from another, and soon children will be working together to move the water along. Likewise, a loft area that has a pulley system with a bucket on it *must* be used by two children, one to put the object in below and one to take it out above. Many of the activities described in subsequent chapters will suggest or require social interaction.

A note of caution: We are not suggesting that solitary play is negative. It is important also to value opportunities for children to work on their projects by themselves. You want to make it possible for many of the activities that you provide to be done alone as well as jointly. Our experience is, however, that too often classrooms undervalue social interaction, and even subvert it by providing too many materials or too few activities in which children must cooperate.

FOSTERING SELF-DIRECTION

Self-directed children have the resources and ability to direct themselves; they don't rely on others to tell them what to do, how to do it, or what not to do. Self-direction is the product of a learning environment designed from the child's perspective. Self-direction is also the product of a good deal of socialization, teaching children that they are able to decide for themselves. Children need models, strategies, and encouragement to be self-directed; in a constructivist classroom, socialization of self-direction is emphasized, and it is reflected in many aspects of the classroom environment.

Let's give some examples of learning environments. In the first one, self-direction is *not* encouraged, because it is not clear, from the child's perspective, what to do or how to do it. Thus children must control themselves, not interacting with the materials *until* someone else, the teacher, tells them what to do and how to do it.

■ Five not-yet-reading kindergarteners are seated around a table with the teacher. At each of their places is a sheet of paper with words and pictures of various objects. There are spaces to check off either "sinks" or "floats," each with a corresponding picture to symbolize an object sinking or floating. In the center of the table is one dishpan filled with water. Sitting next to the pan are the various objects represented on the paper—a rock, a cork, a large plastic foam ball, a tack, and a cotton ball. The teacher holds five pencils in her hand, waiting until she has all the children's attention before she passes them out. She explains that each child will have the opportunity to put one object in the water and that they will all get to mark on their papers whether that object sinks or floats. After the first child, Mary, has put the cork in the water, the children begin to fidget. Cory, whose turn is next, wants to try putting the cork under water with the tack on it, but the teacher repeats again that they each need to try one thing at a time. Without enthusiasm, Cory puts the tack in the water.

Now let's imagine a "similar" activity in a kindergarten classroom where self-direction *is* encouraged.

■ A few kindergarteners are gathered around a table set up with four dishpans filled with water. In the center of the table are objects of similar sizes with varying degrees of density, such as plastic foam balls, rubber balls, and wooden cubes. Peter starts putting objects in the water, leaving them in as he adds others. He runs and gets a popsicle stick off the shelf and pokes the floating objects with it, attempting to sink them. Claire, seeing what Peter has done, goes to the shelf and brings back the container of sticks, picking up a bundle to see if the sticks will float. Watching the sticks float, Claire turns to Peter and exclaims, "Hey, I have an idea!" She goes back to the shelf and brings back some index cards. Carefully placing a card on the floating sticks, she calls to the other children, "I made a boat!" Peter feels around the bottom of his dishpan for a small wooden cube which he had dropped into the water. As he places it on top of the card "boat," he says, "*Now* it floats."

In the first classroom, the teacher's role was to let children know what to do, since the activity really was not self-explanatory. Because the children couldn't read, they needed the teacher to explain about the chart. The rule of putting only one object in the water at a time was necessary only because the teacher had structured the activity that way; the children could not know this was so and, in fact, their initial inclination was to put more than one object in at a time and to use the objects together. Because the children had to "wait their turn," they became bored and distracted. They did not engage in any active experimentation.

In the second classroom, by contrast, the children knew right away what to do. The teacher had decided to put out objects of similar size but different density only because there had been previous experience with the same activity where children had lots more choices of objects to put into the water. This time, the teacher thought it would be interesting to see if the children would make the more direct comparisons among the objects they were given. The children had the opportunity to experiment with other ideas too, though, and Claire's great idea of making a boat introduced other areas of experimentation that the teacher could capitalize on in designing future activities. These children clearly felt free to pursue their own ideas and did so in a very self-directed manner. This self-direction was facilitated by having ample materials, several dishpans, and, most important, the implicit acceptability of taking the initiative and pursuing their own ideas individually.

ENCOURAGING SOCIAL PROBLEM SOLVING

Much of the curriculum we describe in this book focuses on the solving of problems that the child may encounter spontaneously or that may be posed by the given activities and materials. This is because we feel that a large part of what children can gain from an educational experience is lots of support, encouragement, resources, and tools for solving their own problems. In the world that children will grow into, they will be encountering problems that we, today, cannot even imagine. Thus, an "intelligent" person is one who is able to come up with solutions to problems or situations that he or she has not encountered before. Intelligence, as Piaget has defined it, is adaptation.

This is true both in terms of problems that children encounter in the physical world and those they encounter in their social environment. Thus, part of what educational experiences have to offer children is tools for dealing with the inevitable and unpredictable social problems they will encounter as well as opportunities for using those tools in a supportive and facilitative environment. Children must construct those tools through their social experiences, just as they construct any form of knowledge. Yet we don't want them to do this at the expense of others.

A social environment that encourages social problem solving is thus an important component of a constructivist classroom. By devising contexts in which children can encounter conflict and handle it in nonhurtful ways we can clear the way for the positive, creative social interactions that are necessary for conceptual growth and change.

Broadly speaking, there are several keys to doing this. The first is to create a just and a reasonable environment—one where the necessary classroom rules are clearly explained to the children. The first step toward achieving this is for the teacher to ask herself or himself, "What would be the reason for this rule? Is it a reasonable and realistic expectation to have, given the ages of the children I teach?" For example, it is realistic to acknowledge that many preschool-aged children will want to run around inside the classroom. However, it is *not* reasonable to allow children to do this indoors, where they could slip and fall against something or collide with other children. Therefore, "Walking only inside" is a reasonable rule. Next, the teacher must make this rule clear to the children and facilitate their understanding of the reason for it. This may be done by both spontaneous and teacher-planned discussions. An appropriate question will go further in promoting a child's understanding of the rule than a statement. For example, a teacher might stop a child from running, point to the corner of the table, and ask, "What would happen if you slipped when you were running by this table?" During group discussions, children enjoy dictating "rules" to the teacher on a large piece of paper that may be posted, illustrated, and referred to at appropriate times as the need arises.

Another key to setting up the constructivist social context is active intervention—what some have called "guidance" rather than discipline.

Don't miss an opportunity to use conflict between children for educating them—helping them to learn how to handle it. Jump in and give them the support in solving their problems that they may need, particularly very young children. Older children, too, may not have had the experience of using words, explaining feelings, or solving problems either at home or in other educational settings. So don't expect them just to learn *through* interactions; support them in learning how they can handle disagreements. You can also draw on those children who *do* have the capability to help those who don't.

Early in the year or anytime when particular problems arise, it is useful for the whole group to focus on a problem area and on techniques for handling it. One good technique is the use of puppets to replay conflict-generating situations. The children, in discussing the ways the puppet could react, can explore the possible effects of different reactions. This technique is effective because the solutions do not come only from an adult but are elicited from the children.

Another key to creating a "reasonable and just" environment is modeling by the adult. Adults can, through their actions, demonstrate that they too face difficulties and have to solve them. They can show that it is okay not to know what to do, that they can use other people as resources for ideas, and that the resulting possibilities can then be tried out. Modeling is a powerful tool for demonstrating rather than "telling" children about other ways of handling conflict, disagreement, or uncertainty. In Chapter 4, we will be discussing more specifically ways of talking with and questioning children that would be consistent with such modeling. For now, let's look at just one example of how an adult might interact with children who are engaged in a conflict situation.

■ Imagine this conflict with a group of 4-year-olds. Willy and Mark are in one corner of the construction area building a "castle" with large, interlocking blocks. Kathy, who has been watching their progress, clearly wants to join in. Attempting to gain access to their play, Kathy goes to the supply area, makes a "sign," and puts it next to the castle. Willy, misinterpreting her intentions, yells, "Hey, don't wreck our castle!" He crumples up her sign. Hurt by this misinterpretation, Kathy enters the area and knocks down a castle wall. Mark shoves Kathy. The teacher, seeing the last part of this conflict, comes over to the children. "It looks like we have a problem we need to talk about," she says to all three children. Kathy covers her ears and starts to walk away. "Kathy, that makes me frustrated when you do that. We need to use our words now and listen to each other." The teacher continues, "Let's go look at the chart we made last week at group time." The children follow the teacher over to where a dictation is posted, titled "Ways to Solve a Problem if Someone Isn't Nice to You." Willy, who has the chart memorized, blurts out, "Use your words, not your body." Turning to Kathy, the teacher says, "Kathy, can you tell Mark how you felt when he hit you?" "I felt mad," replies Kathy. Mark replies, "But you knocked over our wall and that made me mad." Kathy interrupts, "Yeah, but you wrecked my sign." The teacher then facilitates a discussion by making sure that all the children express their feelings, talk about what made them mad, and together decide what words they could have used.

Later that day, at group time, the teacher uses two puppets who knock down a block building. Clearly the idea is to encourage the children to describe how crumpling someone's sign or wrecking their building makes them feel. Garret jumps up excitedly, saying, "I know, we could get a sample!" Garret runs and gets a piece of paper, draws a mark on it with a marker, crumples it up and tells the group of children, "This is what happened. If this was *your* paper, how would *you* feel?"

This teacher was able to acknowledge the children's feelings, encourage them to come up with their own ideas, and relate the problem to rules generated by the children.

CONSTRAINTS AND HOW TO DEAL WITH THEM

You may encounter certain constraints as you implement a constructivist science curriculum. In order to be effective it is important to consider how to work with, lessen, or eliminate such constraints.

Physical Space and Furniture

Sometimes classrooms are simply not suited to our purposes or they are equipped with the wrong kind of furniture. There are, in fact, very few settings that are *ideal* in every way. Here are a couple of tips for dealing with that typical, imperfect classroom.

First, look carefully at the setting, not critically but positively. What is *good* about the space with which you are obliged to work? Maybe you don't have adequate floor space but you do have wonderful natural lighting with low, child-level windows. Or maybe instead of the large open space you would like you have several smaller rooms connected by doorways. Take what you've got and try to think of what the space you have lends itself to. For example, small spaces may lend themselves to some interesting activities that *have* to be separated from other activities. In one program, for example, a separate small room has become the place for large blocks, rollways, and balls; the sometimes wild activity that takes place there is well confined and more easily supervised by being in a small room.

You could be faced with the alternative problem: space that is too large and too open can be overwhelming and might lend itself to too much running and roaming. This problem is easier to deal with, because a large open space can be partly blocked off and partitioned, whereas a small space cannot always be opened up.

Also, consider ways to change your space. When remodeling is not possible, think about adding a loft or even a platform area that could add diversity to the room.

If it is too difficult to think in terms of the whole space, look at a part of your space. Is there one part of the room that can be more flexible, that can lend itself to more activity, or where a small group of children could engage in a small-scale activity of the sort described in this book? In almost every space, including the most constraining, there is *some* opportunity to move furniture and make space.

Obtaining and Storing Materials

Because some of the materials that lend themselves to constructivist activities are large, odd-shaped, and even "junky," storage can be an issue. If you have access to a closet space, try to use every inch of it, including the ceiling. Raise netting to hold things in on the ceiling, then put storage shelves up to that point. Or, instead of netting, hang big sacks of materials that are not too heavy like tubes or Styrofoam.

The important thing is not to let a dearth of storage space constrain you. Many materials can be obtained as needed, and parents, too, can be resources for storage when necessary. Keep the scope and scale of things reasonable if you find that storage is a major hindrance to implementing constructivist activities.

Constraints from People

Teachers often feel constrained by what other people want and expect of them. This is particularly true in communities and schools where there is an overemphasis on nonconstructivist education. In the following chapters, we shall address issues related to this. For example, we'll discuss the importance of proactively educating other people about the value of constructivist activities. However, the most important factor in dealing with other people is to feel confident about what you are doing in your classroom and why you are doing it.

Without playing down the very real constraints that teachers feel—whether they stem from the physical environment or from other people—we hold that teachers ultimately have to take the responsibility for doing what is right for those who really count: the children in their classrooms. And by focusing attention on children, we can usually find ways around the constraints that may face us initially.

Chapter
4

The Role of the Constructivist Teacher

In a constructivist classroom, the teacher is no longer responsible for "transmitting" knowledge to children. In some ways, this lifts a tremendous burden, since the teacher is no longer *the* source of information, whose presentation makes or breaks the "outcome." But the teacher's responsibilities are in no way reduced in a constructivist classroom; rather, they are shifted. No longer the source of information, the teacher must instead play many different roles. The responsibilities are more numerous and more complex. Because of this, the constructivist teacher's curriculum can *never* be "teacher-proof." The teacher is the core, the orchestrator, the creator of an environment in which learning thrives. This is no easy task. In this chapter, we will elaborate on some of the many roles the constructivist teacher must perform. *All* teachers could and often do view teaching as incorporating these roles. For the constructivist teacher, however, the multi-faceted teaching role is even more important and is different in nature.

Before we elaborate on these many roles, however, let's talk about what it means to be a facilitator of learning rather than a mere transmitter.

A transmitter of knowledge has a clear-cut job—deciding what knowledge to transmit, communicating it to the students, and verifying what they have learned. The facilitator of knowledge construction, on the other hand, has to contend with the fact that many aspects of the learning process are opened up to children's input and are not solely under the control of the teacher. The material to be learned will vary, for example, depending on each child's prior experience, current interests, and level of involvement in the learning encounter. Through their self-directed interactions with an activity, children may alter the activity itself, adding elements, having ideas, and extending it in directions the teacher may not have foreseen.

Although the teacher "loses control" in certain respects in moving toward being a facilitator, he or she still has control and responsibility for a greater part of the child's educational experience. We shall elaborate on these elements in describing how the constructivist teacher performs these numerous roles in the classroom. Briefly, they are as follows:

> *The teacher is a* **presenter**—the teacher must present activities to groups of children, present options to individual children, present ideas to children engaged in ongoing activities. Presenting is different than transmitting because it implies that what is presented is available for children to take or leave.
>
> *The teacher is an* **observer**—in order to present good, facilitative options, to interact appropriately, and to understand the children's interests and knowledge, the teacher must be a constant observer in both informal and formal ways.
>
> *The teacher is a* **question asker** *and* **problem poser**—the teacher must be able to ask questions and pose problems that stimulate theory building without being disruptive to the child. The ability to do this stems from the observation of children as well as an understanding of them.
>
> *The teacher is an* **environment organizer**—the environment must be carefully and clearly organized so that *children* know what to do. Organizing the environment from the child's perspective is an important part of encouraging self-direction.
>
> *The teacher is a* **public relations coordinator**—the understanding and support of many people is critical for the success of any educational effort. And because of the nature of the constructivist classroom, the teacher must clearly and consciously articulate what is going on and why.
>
> *The teacher is a* **documenter** *of children's learning*—meaningful documentation of what children do and what they learn is an important aspect of today's classroom. As we move to more complex curricular goals, simple, quantitative measurements will not capture what children are doing and learning.
>
> *The teacher is a* **theory builder**—in order to be responsive to children, teachers must cultivate their own understanding and interests and maintain their enthusiasm. Nurturing your own capabilities and preventing burnout is important and often overlooked. Just like children, teachers need environments and support for building,

growing, experiencing conflict, changing their ideas, and always being open and interested when ideas don't work as anticipated.

It sounds complicated, and it is. But in the following sections we will elaborate on each of these important roles of the teacher in a constructivist classroom.

THE TEACHER AS A PRESENTER

Teachers must present materials, choices, activities, and options to children throughout the day. Sometimes this "presentation" occurs in large groups and sometimes in small ones. Sometimes presentations require the teacher to talk to the children, to demonstrate, or to elicit ideas through appropriate questioning. In the following three chapters, we have interspersed scenarios describing how teachers might present activities and extend them. Here we will consider some of the general issues related to presentations by teachers.

Presentations by the teacher are important for a number of reasons. First, the way an activity or a set of materials is introduced has the potential to *limit* what the children do subsequently. Let's see how this might happen.

■ When the first-graders come in from recess, each finds a shallow cardboard box lid, construction-paper strips, scissors, and glue placed on his or her desk. The teacher announces that they are going to make marble mazes. She demonstrates to the children how to bend up both ends of a strip of construction paper to glue on the inside of the box lid, making an arch for the marble to roll through. "Here's one that I made to show you," she says as she holds up the maze for the children to see. "Are there any questions?" Alice raises her hand. When she is called on, she asks, "How do we know where to glue the arches?" "That's a good question, Alice," comments the teacher, "Here are some patterns you might want to use." She holds up some dittoed maze patterns sized to fit inside the box lids. "If you are using these patterns you may either glue this in the lid first or copy the patterns directly on your lid. Now let's get started. Remember, watch the amount of glue you use. Too much will take too long to dry. You have 15 minutes to make your maze." For the next 15 minutes the children fold strips according to the teacher's directions and come up to the front to inspect the teacher's model and to take a dittoed maze pattern. Most of the children elected to

In this activity introduction, the children are making predictions about what will happen to the shaving cream if it is put on the vertical sheet of plastic.

glue the dittoed pattern inside the box lid. Those that didn't tried to approximate with varying degrees of accuracy the maze pattern on the ditto sheet.

In the above example, the teacher's presentation limited possibilities. In contrast, the presentation can *open up* possibilities for children.

■ Another first-grade teacher presents the same marble maze activity in the following open-ended manner: "Today, during activity time, one of your choices will be marble maze making. I have some box lids, some paper strips, and some glue. Does anyone have ideas on how to fold the strips so they stand up like this (holding the strip in an arch) for the marbles to roll through?" she asks the children. "I know," volunteers Natasha, "you could fold the ends out." "Or you could fold the ends *in*," adds Nicole. The teacher holds up two strips folded the ways Natasha and Nicole suggest. The teacher then says that a fun maze is sometimes tricky to roll a marble through. "What are some ways to glue on the arches so that it's tricky?" she asks. Marcie suggests cutting the strips in half so the arches are little. Frank's idea is to place the arches crooked and not in a straight line. The teacher finishes the introduction by saying that there are many, many different ways to make these mazes and that the activity will be set up all week long on the project table.

During the week many of the children elected to spend time in this area, some completing one maze, others elaborating on the same maze for several days, and still others creating many different mazes. Many variations of mazes were made and some of the children invented new methods of maze making, such as cutting holes in the box, gluing small pieces of sponge on the box, and making a "tiered" maze out of two cardboard box lids.

The presentation in this case encouraged diverse and creative experimentation. Presentations also occur spontaneously and on a smaller scale. Teachers often have to present ideas, offer additional materials, or facilitate problem solving with smaller groups of children.

■ Emily, 4 years old, has been playing with large interlocking blocks for a while, building horizontal structures. She begins stacking the blocks vertically to create a tall structure. When she has created a structure that is as tall as she can reach, she looks at the remaining blocks on the floor, glances at the top of her structure, and starts to walk away. The teacher, who has been observing Emily, asks her, "Can you think of a way to reach up higher?" Emily stops, looks at a chair, moves it over, and—holding a block in her hand—climbs on the chair to add yet another level to her vertical structure. After a few blocks have been added this way, the structure falls down, ending up still together but horizontal. This time the teacher asks Emily a more direct question: "I wonder how you could use the loft to reach to the top of your building?" Emily says, "I know!" and carefully lifts her structure up so that it is leaning against the loft. Then she climbs to the top of the loft saying, "Hand me a block, please!"

Emily's teacher was able to present ideas to her in ways that facilitated Emily's problem solving.

THE TEACHER AS OBSERVER

All teachers are observers of children, usually in less formal ways in the context of classroom activity. In a constructivist classroom, observation is essential; it is perhaps even the teacher's primary function. This is because observation serves as the basis for everything else that teachers do. Observation gives teachers information about the children's understandings and interests. It provides clues to understand individual children's needs and to help them solve problems. It constitutes the basis for curriculum development. And observation permits the teacher to decide whether and how to interact with children in order to facilitate their activity.

There are many methods of observation that can be adapted for the teacher's use. In this section, we are going to elaborate on two that we have found to be most useful for classroom practice—anecdotal notes and target child observations. You should be aware, however, of the wide range of other methods available, ranging from checklists to the use of videotapes.

Most teachers have had little, if any, training in how to do observations; or, such training may be limited to the more formal techniques such as checklists. For observations to be useful they need to be done carefully, and even the most experienced observers can benefit from practice and feedback. One of the best ways to do this is to practice written observations in a variety of ways. So if you choose to try the methods we describe, *or* if you develop or use others, give yourself time to evolve as an experienced observer and for your observations to evolve in their usefulness to you in the classroom.

Evaluate Your Needs for Observation

Before you incorporate any regular method of observation into your classroom, carefully consider why you need it. Do you have to make occasional evaluations of children in particular curriculum areas? Do you usually have term-end conferences with parents, at which you share information? Is there a curriculum area or even a part of your room which you feel is not working out satisfactorily and that you want to improve? Do you have a lot of parents asking you about their children's writing experiences and an unexpected need for information about what children are doing in literacy? Do you find lots of things happening all of the time that you "just wish" you had captured in writing at the moment they occurred?

Each of these questions could lead to the inclusion of some kind of observation system in your classroom.

Anecdotal Notes

In an active group of children with lots of different things to do, interesting and important things happen throughout the day, in many different contexts. The teacher who schedules a regular time to make observations may find that he or she is missing some significant observations unless there is a

Observation provides the teacher with clues to understanding individual children and to helping them solve their own problems.

system for capturing spontaneous, "unscheduled" wonderful moments. The system of anecdotal observations that we have used can capture these.

"I never have paper when I need it." We have solved this problem by wearing "necklaces" with 3-by 3-inch Post-it™ notes clipped on and a pencil attached. As each note is written it is stuck on the back of the pad. By writing on only one side, we can duplicate the completed notes if we want to share them with parents. We always date each note and at the end of each day file them all in the children's folders, categorized as to whether they deal predominantly with the child's cognition, social interactions, or language. Occasionally we might create a specific category in which to make observations, such as children's spontaneous problem solving, or we might focus on an area that is of concern, like instances of aggression.

In writing the notes, it is important to be as descriptive as possible, assuming that the person reading the note has no "picture" of what is happening. Also, it is very important to be objective and not to interpret the child's behavior as you describe it.

Here are a few examples of "bad" and "good" notes. The first two are not descriptive enough to be useful:

"Mary was playing blocks and counted to ten."
"Paul was busy for ten minutes at the listening center."

Here are two that are more descriptive:

> "Mary stacked ten blocks vertically and counted to ten with one-to-one correspondence, beginning with 'one' and the first block, saying the next number each time she placed another block on the stack."

> "Paul sat at the listening center for approximately ten minutes listening to *The Three Little Pigs* with the headphones on and looking at the book, saying the words out loud and turning the pages at appropriate times."

These next two anecdotal notes are "bad" in the sense of not being objective.

> "Sarah was very enthusiastic painting her piñata. She had fun."

> "Lana was sullen and did not get along well with the other children in the drama corner."

Again, two "better," objective, anecdotal notes:

> "Sarah sang to herself while she painted her pig piñata, laughing each time she added another facial feature."

> "Lana sat in a chair in the drama corner frowning and holding on tightly to the box of 'jewels.' Each time a child approached her requesting the jewels, she turned her head away and did not respond verbally to the request."

Remember, if these notes are to be useful, they must be clear to someone (a parent, a director, a principal) who does not know the context, the child, or the particular event. Also keep in mind that although you may *think* you won't forget the circumstances or the context, in a few weeks you probably will.

The value of anecdotal notes is both in the immediate picture they give you of the children's activities and the cumulative information that you will acquire. You must schedule time at the end of the day or every couple of days to put them in files, so that the task does not become so large that you don't want to do it at all. You must also schedule time periodically to read the accumulated information in the files. We do this both during weekly curriculum planning, when the accumulated notes give us ideas for designing and extending activities, and every 2 to 3 months, so that we can get a picture of each child.

In time, you'll find that your notes focus on some children more than others. To make sure that we take notes on all the children, we tally the number of notes we have on each child in each category as we file. When we notice that we need more notes on some children, we make a special effort to observe that child, perhaps even focusing just on two or three children in one day.

It is very useful to have a body of anecdotal notes for "justifying" things that you do that may not be immediately observable to the untrained eye. For example, if your principal is questioning whether or not the children in your kindergarten classroom are getting enough experience in math activities, you can pull anecdotal notes that illustrate the numerous spontaneous examples of this type of activity.

For example, you might select the following notes:

"Manuel sat at a table with the pattern blocks and spontaneously created a design with all the small squares, triangles, and circles on the left side and all the large squares, triangles, and circles on the right."

"At the snack table today Josie counted out three crackers, three apple slices, and two pieces of cheese for herself. She then did the same for Steve and announced, 'There, now we have the same!' "

"When Jerome and Mike could not agree on who could have the blocks for his airport, Paul suddenly dumped the blocks out and divided them up into two equal piles."

Similarly, if a group of parents in your preschool were concerned that you were not doing enough "prereading" with their children, you could pull out the following notes to illustrate how pervasive literacy and language are throughout the curriculum.

"Billy made a note for his block building like this:

PLZ   D NT   MV

and 'read' it slowly out loud, 'PLEASE DO NOT MOVE!' "

"Chin pointed to the snack sign, which read 'TAKE 3 CRACKERS' and told Fred that the sign said 'Take three crackers only.' "

Anecdotal notes are also a very informative and interesting part of a portfolio which can be the basis for teacher summaries of how children are doing. By sharing anecdotal notes with parents, you can give them a good picture of what goes on in the classroom and what their own child's experiences have been. As we said above, it is very important to maintain the objectivity and nonevaluative nature of the notes, particularly if they are to be shared with parents. While teachers may, after accumulating information from a variety of sources, make evaluations of children ("LaVonne seems to be having some trouble making friends"), the anecdotal notes would provide lots of information about why one would come to that conclusion. It is much more constructive for your own problem solving, and for parents as well, to be very clear about the reasons for evaluative statements.

And observations can sometimes dispel mistaken ideas that you may have formed from more impressionistic information. For example, while a teacher may have the impression that LaVonne is having trouble making friends, closer observation may show that this may not be true. The following notes reveal, rather, that LaVonne is choosing to interact in quieter ways with her peers.

> "In the housekeeping corner, LaVonne nodded yes in response to Mary's question, 'Do you want to be the baby?' She then lay down on the bed with a teddy bear and blanket."

> "LaVonne sat down at the play dough table in front of a big piece of play dough and spontaneously broke it in half and gave half to Douglas. She said, 'Let's make worms,' and both children worked for 5 minutes rolling worms and not talking."

Target-Child Observations

Another useful form of observation is to focus on one child for a more extended period of time, say 10 minutes. One can gain insights into a child's play activities, social interaction, and behavioral styles by concentrating one's attention on everything that child does for one observation period. Some teachers make a habit of gathering such observations and adding them to the child's portfolio; this "forces" you, as the teacher, to zero in on one child at a time. This method is particularly helpful for gaining a better understanding of those children whose behavior is sometimes overlooked— the children who are neither disruptive nor particularly active and therefore may not be the focus of as many notes. There are other teachers who may not have time to observe all their children in this way; they may, instead, use target observations when questions come up about a particular child. Several target observations of LaVonne, for example, could add to our information about what exactly is going on in her social interactions.

We use as our basic technique the method well described in Sylva, Roy, and Painter's book *Childwatching at Playgroup and Nursery School* (1980). This involves a 10-minute running record of one child, in which we write down everything the child does without making evaluative statements. We also try to capture as much language as possible. It is helpful to divide the paper in half across its width, with the description of the child's action on the left and the corresponding language on the right. Some teachers prefer also to divide the paper into ten sections lengthwise and record the child minute by minute. Abbreviations such as "T" for teacher and "C" for child are helpful, as well as the symbol → to convey "speaks to." For example, "C → T hello" would mean the child said "hello" to the teacher. (In *Childwatching*, the authors go on to describe how to code the resulting observations; for teaching purposes, this is not necessary.)

As with anecdotal notes, you'll find that you get better and better at writing clear, useful target-child observations. Try to keep in mind the need to write them for an audience that is not seeing what you see. At a later date, this observation will be *all* you have as a record of what you saw; therefore don't make assumptions about what the reader of the observation may know about the room, the children, or the curriculum.

It might seem that it would be hard to fit in the time to do target-child observations. You'll have to plan carefully in order to include these times in your schedule. Good times might be when you have sufficient staff (an aide or parents) available or when the children are out on the playground. You will find that the effort is worthwhile.

THE TEACHER AS QUESTION ASKER

One of the most important links between the teacher as a facilitator of knowledge and the child as experimenter is the teacher's questioning strategies. The types of questions asked and their timing can determine whether the child pursues his or her experimentation. It can also determine whether the child explores problems at a more complex level. In response to a question, the child may change his or her focus to respond to a teacher's need or perhaps even leave an activity altogether.

Let's examine the following scenario to see how a teacher's questioning affects the child's learning encounter:

■ Amanda stands at the water table with a quart container in one hand and a small cup in another. She begins scooping up water in the cup and filling the quart container to the point of overflow. Amanda sticks her hand in the quart container, causing the water to overflow. Catching the overflow with her other hand, Amanda keeps pouring water into the quart container.

Amanda is engaged in some interesting experimentation with overflow, surface tension, and displacement here. There are various ways a teacher could or could not intervene in Amanda's activity. First, consider this teacher/child interaction:

■ Upon observing Amanda, the teacher asks her, "How many cups does it take to fill up the big container? Can you count them for me?" At this point Amanda stops her experimenting, empties the quart container, and begins filling it up and counting. When her "task" is done, she leaves the water table.

What has happened here as a result of the teacher's questioning? Besides being guilty of not observing clearly what the focus of Amanda's

The role of the teacher is to facilitate and encourage children's theory building.

experimentation was and not facilitating it further if necessary, the teacher interrupted Amanda's theory building for a teacher-directed task.

Now let's consider a type of teacher interaction more consistent with a constructivist approach. Perhaps it might be appropriate for the teacher *not* to interrupt Amanda with questions. She appears, after all, to be pursuing her own questions with concentration. Or, perhaps if, after a few minutes, Amanda stops, it might then be appropriate for the teacher to ask some questions. These questions *must be consistent* with what the child is actually exploring. For example, the teacher can see by observing Amanda plunge her hand in the water and catch the overflow that Amanda is exploring water displacement. The teacher could ask, "I wonder what would happen if you put those big rocks in the big container?" Or if the teacher observed that the focus of Amanda's experimentation was overflow and comparisons of volume, the teacher might ask, "What would happen if you poured water from the big container into the little container?"

The important concepts to remember are that the teacher's questioning strategies or, to use George Forman's term, the teacher's "point of entry," must:

1. Support the constructivist view of the child as scientist and theory builder.

2. Be based upon observation of the child.

Some simple guidelines will be useful in deciding how to phrase questions. First, think in terms of questions as being either "open-ended" or "closed-ended." There are many possible responses to an open-ended question as opposed to a closed-ended question, which usually calls for only one answer or even just a simple yes or no. Examples of closed-ended questions are "What color is it?" or "Is the brown box bigger than the ball?" Generally, open-ended questions are desirable and consistent with a constructivist approach. In some circumstances, however, closed-ended questions are necessary. For example, if, after a lengthy discussion about the danger of waving sticks around, a child still persists in doing so, you might ask, "Is it a choice to wave a pointy stick around?"

Second, the goals of your questions will usually be to encourage problem solving, perspective taking, and/or consideration of feelings. And third, some key phrases can be helpful to "plug into" different situations. Some of these phrases to remember are: "Can you think of a way to . . . ?" "Do you have some ideas about . . . ?" "How do you feel about . . . ?" "What would happen if . . . ?" These suggestions are also useful in planning activity introductions, as discussed previously in this chapter, and in the sample activity sections of Chapters 5, 6, and 7.

The following are some different situations in which the teacher will need to make decisions about questioning strategies.

When the Child Directly Asks the Teacher for Help

Even when a child comes to you with a request for assistance or a question, you can, by responding with the right question, challenge the child. For example, if a child asks, "How can I reach up high?" the teacher might respond with, "Do you have some ideas about what is in this room that you could use?" Or if a group of children come running in from outside and announce that the ball is stuck in the tree, you could ask, "Can you think of some different ways to get the ball down?"

Crisis Intervention

Crisis intervention is usually necessitated by conflict, safety considerations, or unreasonable messes. In either case, once the immediate danger or problem is past and the child or children are focused on you, the goals of your questions should be to generate perspective taking, leading to acceptable solutions. Questions such as "How did that make you feel?" or "What would happen if . . . ?" generate perspective taking and an understanding of the situation. Questions such as "What are some ways to . . . ?" or "How can we . . . ?" can generate solutions to problems. For example, if a group of children were playing so wildly at the sand table that some sand was flying through the air, the teacher might ask, after getting their attention and stopping the sand throwing, "What would happen if the sand got too near someone's face?" The children's responses could be followed by another question, such as "What are some safer ways to play with the sand?" As we said earlier, once in a while a more direct, closed-ended question *is* appropriate. For example, "Is it a choice to use sand in a way that might hurt someone?" This question focuses the children on the problem that needs resolution.

Encouraging Children to Explore Further

A careful observer of children can grasp the particular challenge or focus of the child's learning encounter, and ask an appropriate question at an appropriate time to stimulate the child further. The first question the teacher should ask is, "Will my question to the child enhance or detract from the child's exploration?" Usually an actively engaged child in a well-designed environment will benefit most by *not* being distracted by questions. However, there are some guidelines to use as you decide what to do. If the child pauses or appears to be about to leave, you may be able, by asking a question, to encourage the child to probe at a more complex level. For example, consider a child who has been creating a vertical structure with large interlocking blocks. The child stops when she cannot reach any further. The teacher might ask, "What are some ways you could reach higher?" In this way the child is encouraged to keep building. If the child cannot come up with a solution, the teacher could change the questioning strategy and ask a more direct question, such as "How could you use this chair to reach higher?"

An important principle to remember is to start with the most open-ended question possible and only ask a more directed question if it seems necessary. Sometimes, when a child reaches a frustrating dead end, a teacher will have to employ this principle and consider the continuum between open- and closed-ended questioning. An appropriate question can alleviate a child's frustration by offering some ideas yet still encourage the child to continue exploring without making direct suggestions.

A last principle to consider is nonverbal questioning. Sometimes a teacher can ask a question without using words simply by placing a helpful material within the child's view. For example, the child building the tower with interlocking blocks might benefit if a chair were placed nearby without a word. Or perhaps if the teacher, who is holding the structure for the child, leaned it against the loft, the child might realize (all by herself!) that she could climb up to the loft and add more blocks.

Asking questions, like observing children, becomes easier with practice. You may sometimes be surprised when children begin to ask *you* open-ended questions, such as "What do you think might happen if I let go?" Or you may hear children spontaneously asking each other open-ended questions.

THE TEACHER AS ENVIRONMENT ORGANIZER

Because a constructivist approach to education is based on the belief that children learn by interacting with their environment, the teacher must see the physical environment of the classroom as an extension of his or her role. In Chapter 3 we discussed the physical environment in terms of constructivist goals for children. The principles discussed in that chapter—such as easy traffic flow, flexibility, accessibility, reciprocity, self-direction, and social interaction—all have an impact on the role of the teacher. These principles create a self-explanatory environment that tends to direct children's behavior more than the teacher does.

For example, a classroom with a well-planned traffic-flow pattern eliminates the need for the children to be reminded to walk or move quietly through quiet areas. A flexible environment gives children the message that they can change their environment themselves, rather than asking the teacher or waiting for the teacher to do it for them. A teacher who plans the environment and considers reciprocity or the location of learning areas reduces the need to monitor disruptive behavior or to tell children that certain combinations are not allowed. When materials are readily accessible to the children, their self-direction is encouraged. An environment set up to encourage interaction between children also facilitates cooperation, so that children help each other rather than always depending on the teacher.

In an environment where the transmission of information was important, the activities would be more specific and would require instruction from the teacher. However, in a constructivist classroom, the environment must be designed to allow for many different ways of interacting with the materials. It is, therefore, the task of the teacher as organizer of the environment to consider carefully how children might be expected to use the materials. This can eliminate the need for unnecessary intervention. For example, if you set out paint on a cookie sheet, expect that preschoolers will probably rub their hands around in it. If you do not want this to happen, then put the paint in small containers. Or if you plan an activity in which children throw objects through holes, place a backdrop behind the activity, so you won't have to tell the children to watch out for flying objects.

Even a matter as simple as where to place trash cans should be considered by the teacher. Small trash cans placed around the room, instead of a single large one, can give children the message that they are expected to clean up. Signs with pictures as well as words help direct even nonreaders and also create a literacy-rich environment.

The challenge for the teacher as an environment organizer is to plan carefully, observe the results of the plan, and then evaluate its effectiveness and alter the environment as necessary.

THE TEACHER AS PUBLIC RELATIONS MANAGER

The goals in a constructivist classroom are more diverse, rich, and complex than those in a classroom where "outcomes" can be easily measured (e.g., learning to write the numerals from 1 to 10). Therefore, the teacher's job of explaining to parents, colleagues, and administrators what children are learning is a more challenging one. It is in the best interest of the child that support and encouragement for theory building come from all levels, not just from the classroom teacher. The foundation, therefore, for being an effective manager of public relations and a good teacher is to understand the "why" and the "what" of your perspective on teaching and learning and to be able to articulate these with confidence, clarity, and enthusiasm.

It is a wise public relations strategy to anticipate this task of explaining goals and philosophy and to plan ways of doing so *before* you are asked about it. There are both formal and informal ways of doing this. Orientations with slides or videotapes of the children can be very effective. Newsletters with explanations of why certain activities are planned can be complemented with articles from books and journals. The publications of the National Association for the Education of Young Children—including articles from the journal *Young Children*, brochures, videotapes, books, and other resources—can help you in this task. Give parents suggestions about home activities or attitudes that will complement classroom activities.

Encourage visits to your classroom, too. When parents, teachers, or administrators visit your classroom, first take the time to explain what they might expect to see and the significance of it. This can cause the "just playing" that they observe to take on new meaning. During your discussion about your educational approach, refer to books you've read or workshops that you've attended.

The children can be your greatest resource in communicating with others. Have samples of the children's dictations, signs, and pictures around the room. Label the room with dictations from the children about what goes on in each area. Complement these signs with your own explanations of why you do what you do. Have children take the visitors on a tour of the room.

Whichever of these suggestions you are able to incorporate into your teaching schedule, the most important public relations tool is *you*. You must understand yourself why you are teaching in a constructivist manner and what that means.

THE TEACHER AS DOCUMENTER OF CHILDREN'S LEARNING

Teachers must be responsible for compiling, organizing, using, and sharing information about what children are doing and learning while they are in educational settings. In the constructivist classroom, this role is no less important than in the traditional classroom. After all, parents and administrators will, in any context, have legitimate interests in knowing what is going on and what children are doing and learning. Accountability in a constructivist classroom requires that the teacher be adept at collecting and organizing a more diverse range of "evidence." While traditional approaches may focus on standardized tests and quantifiable measures of learning, the constructivist teacher will collect a diverse array of evidence of learning, including writing samples, observations, pictures, signs, artwork, and written teacher comments. An approach consistent with constructivism is the development of portfolios, which can incorporate numerous sources of information relating to an individual student's performance.

The role of documenting students' work goes hand in hand with the teacher's other roles. In the course of observing children, teachers will collect a lot of information that helps them determine what children know and what they are interested in. In the course of explaining to others, they will marshall all evidence at hand, sharing with parents and administrators what is going on and why. And while implementing the curriculum, teachers will also get many ideas and much information from the children about what they understand and what they need to work on.

THE TEACHER AS THEORY BUILDER

This role is an umbrella for all the other roles. Without allowing the time and energy for your own professional and personal development, you will not be able to fulfill all the other roles effectively and enthusiastically. We have, therefore, devoted the final chapter of this book to considering the teacher as a theory builder, just as the rest of the book focuses on the child as a theory builder.

PART THREE

Constructivist Science

Chapter
5

How Can I Make It Move? Constructivist Physics

One of the most immediate, visible, and comprehensible ways in which children can experiment with the physical world involves the movement of objects. Pushing, sliding, rolling, tilting, throwing; using balls, cubes, wheels, marbles, paint, water; moving things through the air, on water, over bumps . . . the possibilities are endlessly fascinating to children. Consider the following activities, using the same materials, first in a preschool, and then in a first grade classroom. Here is the first:

> ■ Sarah, who is 4 years old, walks over to a large round disk that is on the floor in the middle of the carpeted construction area. The disk is about a foot in diameter, and five or six marbles are resting on it, held by a small lip. She touches one side of it gingerly with her foot. It wobbles back and forth, causing the marbles on it to roll toward her foot and then away again. She laughs and calls to another child, "Hey Mary, come see the marbles tickle my toes!" Mary responds to the invitation and steps on the other side of the disk carefully with one foot. The girls synchronize the wobbling of the disk with their feet, laughing every time the marbles touch their toes.

Here is the second:

> ■ In the first-grade classroom, the disk is sitting on a table. Amanda, who is 7 years old, sits down at the table and rocks the disk from left to right, her eyes following the movement of the marbles. After a few minutes, she carefully tips the disk and moves it to roll all the marbles toward her. She then experiments with tipping the disk to make the marbles move directly away from her. Next Amanda slowly tips the disk, causing the marbles to roll in a circular motion around the disk. She stops every now and then to reverse the direction of the motion.

We are calling these activities *physics*, the physics that young children can understand.

Grownup scientists define physics as "the science of matter and energy and of interactions between the two." This includes a large domain of knowledge about the physical world, and a good deal of it is inaccessible to young children. Our task is to try to look at those parts of physics that are comprehensible to young children.

In this area more than the other two science domains we will be dealing with, we have some good groundwork that has been laid by other constructivist early childhood educators; much of what we will describe relies heavily on their work. The work of Constance Kamii and Rheta DeVries on physical knowledge (Kamii & DeVries, 1978) and that of George Forman and his colleagues (Forman & Kuschner, 1984; Forman & Hill, 1984) on applying Piaget to early childhood education forms the foundation for our understanding of what is appropriate and possible for young children in approaching this often complex physical world.

THEORETICAL BACKGROUND

The Work of Constance Kamii and Rheta DeVries

Kamii and DeVries (1978) define "physical knowledge" as being constructed through "the child's *action* on objects and his *observation* of the object's reaction." They describe two categories of physical knowledge. In the first, which has to do with the movement of objects, the emphasis is on the child's *actions*. An example of an activity that falls into this category would be children's experiments in rolling balls down inclines that can be varied in height by the children. In the second category, which has to do with changes in objects, the emphasis is on the child's *observations*. An example of an activity falling into this category would be the cooking of play dough batter, and the careful observation of how the substance changes as it becomes hotter. Many activities involve both changes in objects and in their movement.

In both cases, there are four criteria to be targetted in maximizing the child's ability to observe the effects of his or her actions on objects. Because we shall be relying on these criteria in subsequent activities, both in this and in later chapters, they will be discussed in some depth. First, the child must be able to produce the movement by his or her own actions. The connection between what the child does and how the object responds must be as direct as possible. If we think of these criteria as falling on a continuum, we would consider a demonstration by another person as falling at one end, where the connection between the child's action and the object's movement is *not* direct. Somewhere in the middle would be an example given by Kamii and DeVries, in which the child is moving an object with a magnet—magnetism being the intervening force that is moving the object. At the other end of the continuum would be a child pushing a ball. Here the connection is direct.

The second criterion for a good activity promoting physical knowledge is that the child must be able to vary his or her actions. Kamii and DeVries say that this is important, because without this ability, the child cannot significantly affect the outcome. For example, in rolling a ball down an incline without being able to vary the height of the incline or vary the type of ball, the child can only reproduce the same action over and over and cannot explore the effects of his or her actions on the outcome. We would say, further, that it is only the possibility of variation that makes experimentation of any kind possible.

The third criterion is that the reaction of the object must be observable. Again, unless the reaction is observable, the child has no way of constructing a correspondence between his or her actions and the resulting effect on the object. This criterion is important because many phenomena in the physical world and many things that we present to children and expect them to understand involve invisible reactions, either because of the speed at which objects move or because the phenomenon is only indirectly observable (for example, gravity, magnetism, or electrical force).

The fourth and final criterion is that the reaction of the object must be immediate. Kamii and DeVries talk about this in terms of the child's ability to establish the correspondence between his or her actions and the reactions of the objects. A time lag in reaction makes that correspondence difficult for the child to construct, so that other explanations for the reaction may be constructed instead. For example, the growing of plants often involves a time lag between planting and sprouting. Upon seeing the grown plant, young children may not make a connection between their actions (preparing the soil, putting the seed in the ground, watering it) and the resulting plant. (This is not to say, however, that these "slow-change" nature experiences are not important for the young child. In Chapter 7, we shall discuss ways to make changes in the natural world more observable for the young child.)

The use of these criteria as guidelines is very helpful from a curricular point of view because each of them suggests ways that activities can be designed and presented to maximize their effectiveness as "good" physical knowledge activities. First, the child's activity must be primary, and the activity must not be a passive or vicarious one (such as an activity that is described in a textbook or on a film). Second, the activity must provide the child with ways that his or her actions can be varied—in other words, experimentation must be possible, and it must be possible from the *child's* point of view, so that the child can establish a correspondence between his or her actions and the reaction of the objects or materials. Too many variables might prove confusing. Think, for example, of an activity where a child can vary not only the size, shape, and weight of the ball being rolled down an incline but also the height of the incline itself. What happens when the child changes the ball and the incline at the same time? The two variables could become confounded and the child could misinterpret cause and effect. We shall discuss this issue in greater depth later on. There may, in fact, be times when you will want to give children opportunities to experiment with *many* variables at the same time. There may be other times when you want to make the variables clearer to the child. The criteria regarding the observability and immediacy of the object's reaction have clear implications for the types of materials to use. Materials such as clear hard plastics which maximize visibility would be desirable. For example, imagine children moving a ball through a clear plastic tube by tilting the tube to various degrees. Here children are able to observe the relationship of the different inclines to the movement of the ball, as they could not do with a cardboard tube in which the balls go in one end and out the other without the changes in speed being clearly visible. Further, activities in which the pathways of the object's movement are clear, where motion is represented, would also better meet this criterion. We will be discussing the issue of representing motion more as we elaborate on the work of George Forman.

The Work of George Forman

The work of George Forman and his colleagues (Forman & Hill, 1984; Forman & Kuschner, 1984) both complements and extends the work of Kamii and DeVries in a number of ways that are integral to the approach presented here. In this section, we shall present those elements that are crucial to the orientation and organization of this book. Those elements are *transformation* and *representation*.

Transformation   According to Forman and Kuschner, a focus on transformation, on the processes of change, distinguishes the constructivist perspective from other views of intellectual development. Knowledge is not a copy of external reality, passively taken in by the child, but is a "construct of the mind." The child does not learn by making increasingly precise discriminations, noting differences between objects, but rather by changing those objects, by acting on them. "According to Piaget, knowledge develops through learning how objects move, how they change position and shape, and how they change in their relation to themselves and other objects" (Forman & Kuschner, 1984, p. 52).

Forman and his colleagues argue that the major intellectual developments all involve understanding transformation. For example, the infant's first understanding of an object involves what Piaget calls the "object concept," which is the realization that an object remains the same even when it looks different—when it changes position or is seen from another angle. Only experiences with objects moving in space, being transformed in appearance, will lead to the construction of the object concept. Thus, the focus in early childhood education should be on the facilitation of transformational thinking. One key to such a focus is the child's *representation* of transformation.

Representation   When you look at the experiences that Forman and his colleagues have developed to facilitate "transformational thinking," many of them emphasize activities and materials that *represent* transformations, whether it be the representation of movement or the representation of change within an object. For example, a spool that rolls down an incline and leaves a blip of paint (a "blip spool") is leaving a track that represents the pathway of the spool's motion. Forman would contend that such representation is facilitating the child's construction of the process of movement from point A (the top of the incline) to point B (the bottom of the incline). Because young children tend to focus on static images—the beginning and end points of change—such materials would encourage the transition to focusing on the transformation and not the static points.

What is the goal of such activities and materials? It is assumed that they will ultimately facilitate the child's representational thinking, that will, then, incorporate the representation of movement and change. This is why the activities Forman, Kuschner, and Hill describe range from those that unitize motion, such as the blip spool which leaves a track of paint as it

Powdered tempera and shaving cream on hard plastic provide clear, visual representations of the movement of the child's hand.

rolls, to those that represent the continuum of change, such as the "swinging sand pendulum," a plastic bottle filled with sand that has a hole in it and releases the sand as it moves.

Forman and his colleagues use a number of catch phrases that are useful in developing activities to represent transformation. "Down with dichotomies" refers to the need to "fight against" the child's tendency to put things into two categories (open/shut; in/out). Similarly, "placing static states along a continuum of change" refers to trying to get children to see the connectedness along a continuum instead of, say, the beginning and end. These slogans can guide us in our curriculum development. When we consider the process that the child is engaged in, Forman's ideas can help us to explore how we can make those processes more visible—more accessible—to the child.

Let's take an example from Kamii and DeVries and discuss how Forman's ideas extend their work. Kamii and DeVries describe, in some depth, the game of target ball. In target ball, children are given materials for the construction of targets as well as balls to aim at their constructions. This activity incorporates lots of opportunities for experimentation, and the teacher can modify it depending on what aspect of the situation might be the focus. For example, you could provide an extensive array of materials with which children could construct targets. This might focus the children on the width and height of the target as variables affecting success. Or, you could design the activity to encourage them to vary their distance from the target.

Consider this child's interactions with target ball:

■ Emma stands at the line marked on the floor with masking tape, looking at the tower standing 5 feet away that she has made out of empty milk cartons. She kneels down on the floor and swings her arm with the ball in her hand as if to aim at the target. Finally, she releases the ball, rolling it to the left of the target. She tries again, this time rolling the ball too far to the right. After several unsuccessful attempts to knock down her target, Emma gets up and rearranges the milk cartons, this time placing them next to each other in a row. Kneeling on the line, Emma rolls the ball, and, as she hits the center carton, exclaims, "I did it!"

Emma is having a great time and lots of good experimentation is going on. Thinking of Kamii's previously described criteria, though, the observability of the ball's movement through space may be limited. Emma can observe point A, her throw, and point B, where the ball hits the target, but the movement of the ball is too fast for her to track very well. Although Emma *may* end up experimenting with how her swing affects the path of the ball, Forman would focus on ways to extend the activity that would allow the track of the ball to be represented and thus more readily experimented with.

One way to extend the activity to increase the representation of movement is to make it a pendulum target-ball activity, with a swinging sand pendulum. Here the child swings a pendulum that releases sand, and the pathway of the pendulum leaves a mark. This would give Emma more information about the path from point A (release) to point B (the target), so that she could experiment with lots of things, including varying that path.

In both activities, Emma would be engaged in active experimentation and ultimately could construct an understanding of the correspondence between her actions, the actions of the ball, and the resulting reaction of the target. By introducing ways to represent motion, the activity can be extended, with the possibility for increasing experimentation that focuses on the representation of the trajectory.

We say it is a *possibility* that Emma's experimentation with the trajectory will be increased, because, in fact, we do not know if it will be. There has been little research on the effects of the different types of activities on the nature of children's play, much less the farther-reaching effects on their theories about the physical world. Much of Forman's work, and Kamii's as well, is based on theoretical reasoning. And because there is much we do not know about the effects of different activities on children's actual thinking about phenomena they are experimenting with, we feel it is important to provide a wide range of activities, carefully observing the children engaged in them to ensure that they are appropriate and intriguing.

Variation and Complexity in Activity Design

One of the issues that comes up over and over as you begin to apply constructivist ideas is *how much variation to design into an activity.*

Go back to the beginning of the chapter and think about the marbles rolling on a tiltable surface. What are the possible variables here that children could experiment with?

One variable is the degree of the surface's tilt. When it is tilted up high, the marble rolls quickly; when it is tilted just a little, the marble rolls more slowly; and if it is tilted back and forth, the marble can roll first one way and then another. The child has the opportunity, through experimentation, to establish the correspondence between his or her actions, the movement of the surface, and the movement of the ball across that surface.

For some children of some ages, this might not be very challenging. Seven-year-old Amanda, playing with this activity, may already have constructed her ideas about the relation between the degree of tilt and the ball speed. "Playing" with this relation might lose its appeal fairly quickly.

To increase the complexity of the experimentation that is possible, we could design the activity to incorporate other variables besides the tilting surface. We could provide different kinds of objects to move on the surface. Then, the children could experiment with how large objects move as opposed to smaller objects, or how balls move as opposed to cubes. You could also offer a series of balls of different sizes, so that children could experiment with the "continuum of change," to use George Forman's term.

We could go even further. We could introduce different ways of propelling an object across a tiltable surface—blowing through a straw, using a bellows, pushing with a stick. You can begin to see that at some point the variables would probably become too complex for children to *really* engage in any kind of experimentation. There would be just too many things to keep track of, and even adults have trouble keeping track of many variables. For example, 3-year-old Julia was involved in the following activity, which probably had too many possibilities:

■ A maze made of wooden unit blocks has been set up on the table. In one box next to the table are balls of various sizes and weights, ranging from a marble to a large foam ball about 5 inches in diameter. Another box contains various objects intended for the children to use in propelling the balls through the maze, such as straws, a paintbrush, feathers of different sizes, toothpicks, and a bellows. Three-year-old Julia selects the large foam ball, which she sets in one corner of the maze where there is a large open area. She attempts to move it gently with her hand to another part of the maze, but the ball gets stuck between two blocks. Next, Julia picks up the straw and blows on the ball, which does not budge. She tries to shove it with a paintbrush and moves the ball but also makes the blocks topple down. She picks up a feather, blows on it with the straw, and walks away.

In this activity, Julia was *not* able to separate how the differences in the balls' size and weight affected their movement from the way the characteristics of the different objects pushing the balls affected their movement. There were simply too many interacting variables for a 3-year-old to consider. You could either offer Julia balls of only one size and several objects to push the balls with, or you could offer her only one type of object for pushing, with balls of several sizes. This would allow her the opportunity to experiment with the effects of the differences and to build some theories about them.

There is no magic rule for determining the complexity of variables that will work. There are a couple of guidelines that we can share with you, though, to help you decide on the complexity of an activity.

First, think about the ages of your children. Older children can handle more complexity than younger ones. But, it still depends on the nature of experimentation.

So, second, think about the particular variables and how much experience children would have had with them. Rolling balls, for example, is probably a more common experience for children than propelling floating plastic balls in water with a straw.

Third, start fairly simply and build in complexity. Usually children have to become accustomed to an activity before they engage in experimentation; simpler activities would still be intriguing even if the children were familiar with the variations embodied in them.

Finally, *watch* the children and use your observations as the guide to whether or not, and how, to increase the complexity of the activity. If children are staying with a fairly simple activity for long periods, if they are engrossed and involved, then changing it may not be necessary at that point. If children are not choosing an activity or if they "do it" and then leave, it is probably too simple.

In the right environment it is often possible for children to introduce complexity themselves, as they become ready. If, for example, they have easy access to other materials and it is *okay* for them to bring in other elements, then unchallenged children will challenge themselves by introducing other elements. And if you observe this happening, it can signal you to encourage it by setting out the right materials or by changing the activity to accommodate the children's great ideas.

■ In this kindergarten, for example, children are encouraged to extend and expand the activities. In one corner of the classroom, the teacher has set up some ramps supported by large hollow blocks and has provided baskets of assorted objects such as cars, balls, and tubes to roll down the ramps. Additional blocks have been set out should the children want to vary the incline. After experimenting for 5 minutes with rolling different objects down the incline, Mary places another large block under the high end of one of the inclines to make a steeper ramp. She places a car at the top of the ramp and releases it. It moves down so rapidly that it hits a block on the floor, bounces up, and lands in one of the baskets. Mary is both surprised and delighted with this result. She quickly dumps out the baskets and places them at the foot of the incline. In the process, she knocks the ramp over. After Mary attempts unsuccessfully to rebuild the ramp, the teacher sets down some masking tape and plastic crates next to Mary, asking, "Can you think of a way to use these to make your ramp stay steep?"

Mary's "extension" here, involving making the incline steeper than the materials laid out were able to accommodate, was facilitated by the teacher, who saw her frustration in trying to act on her ideas. This observant teacher might follow up the next day with an activity where the materials can *easily* make very steep ramps, like a ladder on the wall to be used with "lipped" boards that can safely lean and make high and steep ramps.

So as you can see, there are numerous ways that the issue of complexity and variation in an activity can be considered, both in terms of planning for and supporting children's experimentation.

Categories of Physics Activities

In this section, we shall give examples of children engaged in physics activities in three general categories. Overall, these categories progress from less to more complex; but within each category, the issue of the number of variables, discussed above, applies. And while these categories are not intended to prescribe the curriculum sequence rigidly, we have found it helpful to use them in considering the progression of activities that we offer to children.

The categories are logical progressions in how children come to understand and increasingly conceptualize the movement of objects. In the first category, objects are to be moved, and activities are designed to provide opportunities for actions on objects that make them move. In the second category, the movement of objects is more directed, with emphasis on movement toward something (aiming) and direction of movement. In the third category, the connection between the movement of objects and the actions of the child is the focus, and the representation of the object's movement is of interest. Under each category, we shall provide descriptions of children at two different levels engaged in sample activities.

Category 1: Moving Objects   In a preschool classroom, children are experimenting with the effect that changing the incline has on the movement of the ball. Notice how Erik introduces a different type of movement to the activity.

■ Suspended from the ceiling at "kid level" is a large hoop covered with clear plastic sheeting. A small cardboard rim is taped around the hoop's perimeter so that the balls placed on it will not fall off. Brittany pushes the hoop so that it swings back and forth. After several pushes, Erik joins Brittany and twists the hoop so that, as it unwinds, it reverses direction. Both children watch the balls move around the hoop and laugh. Erik gets under the hoop and says, "I see the balls!" He accidentally bumps the bottom of the hoop with his head and the balls bounce up. Brittany laughs and says, "Bouncy balls, bouncy balls!" Erik purposely bumps his head on the underpart of the hoop, bouncing the balls up and moving them around. Both children laugh and continue their chanting.

In the following activity in a second-grade classroom, the overhead projector highlights the movement of the liquid as it responds to the children's actions.

■ A clear pan of colored water with drops of oil in it is placed on an overhead projector. Seven-year-old Peter cautiously tips the pan and observes the movement and level of the liquid in the pan without looking up at the screen. Minalee notices the movement of the liquid projected on the screen and exclaims, "Look at the moving circles!" Peter glances up and begins tipping the pan in different directions while glancing at the screen to see the effects of his actions.

When pulling "down" means the sand goes "up," these children focus on directing movement.

Category 2: Directing Movement In this activity, preschool children direct the movement of the boat with a stick.

■ The water table is filled with 3 inches of water and unit blocks are placed parallel lengthwise in the table with gaps every so often. Avery, a 4-year-old, places a small boat in the water and moves it through the "route" with a popsicle stick. When it gets to the end, it bumps the side of the water table.

While Avery moves to that end of the water table to push her boat back, John announces, "I'm going to make bumpers for you!" He places some small blocks every so often. Avery pushes the boat to the blocks and laughs as she flies the boat over each "bumper," continuing through the maze.

In a first-grade classroom, the children direct the movement of marbles by creating and changing rollways, experimenting with inclines.

■ A group of the children have been given a box of wrapping-paper and paper-towel tubes cut in half lengthwise, some tape, some marbles, and some small blocks. The teacher suggests that

These children have placed a stack of blocks where they predict the ball pendulum will swing.

they think of different ways to make the marbles move. The children immediately begin constructing "roller-coaster" ramps by balancing the tubes, held secure by tape, on blocks of varying sizes. They experiment with moving the marbles up inclines and down inclines and around corners. Darrell, after successfully moving a marble down a long ramp and around an angle, props a small tube upright at the end of the ramp. He then takes one of the larger marbles and releases it at the top of his ramp. The marble rolls down and around and knocks the tubes over. Darrell exclaims, "Look!" Several children turn from their constructions to watch Darrell's target practice and then set about to build targets of their own.

Category 3: Representing Motion   In this preschool classroom, objects dipped in paint focus the children's attention on where the objects move.

■ A large piece of sturdy cardboard (approximately 6 feet by 6 feet) is covered with butcher paper and leaned against a table securely so that it is at an incline. A cookie sheet—filled partly with paint and various objects such as small cars, spools, and balls—is set on the table. Blocks are set on the floor at the foot of the cardboard to prevent the objects rolled down it from continuing out into the room. Mandy and Justin, both 3 years old, approach the activity, pick up cars, and roll them down the cardboard, laughing when the cars bump into the blocks. After a few minutes of this, Mandy notices the paint and gingerly dips her finger into it. Then she runs her finger along the cardboard, making a line. She looks at the tray of objects, selects a car, and dips the wheels in the paint. Justin picks up one of the spools and rolls it around in the paint. "One . . . two . . . three . . . go!" shouts Mandy, releasing the car at the top of the incline. "Hey, I made a road," she says, surprised. Justin releases his spool and asks Mandy to find his road. Both children continue dipping their objects in the paint and rolling them down the incline as they tell each other, laughing, "See my road!"

In the following example, one child in a first-grade classroom is challenged to follow another's movements, as highlighted by a flashlight.

■ Bob, aged 7, and Maya, 6, stand on opposite sides of a large piece of vertical clear plastic to which plain newsprint is fastened. Maya holds a flashlight and moves it around quickly as Bob tries to track it on the paper with a marking pen. Maya giggles and says, "Let's pretend this is a magic flashlight and it makes you draw things!" She then makes slow, circular, connected movements with the flashlight. "Hey," says Bob, "The magic is drawing bouncy springs!" Maya begins to move the flashlight faster and faster as Bob tries to follow with the marking pen. Finally he says, "You used the magic up, now it's my turn to try it." They then switch places and continue their game.

Sources of Ideas: How to Find Activities

Once you start looking for activities and materials that relate to the movement of objects, you begin to see them everywhere you look. Lots of commercial games and materials lend themselves well to object movement activities. And many curriculum resource books describe activities that can serve, or be modified to serve, our purposes. In this section, we'll offer a few ideas on what to look for and also some guidelines for choosing activities. The recommended materials—along with those described in other chapters—are also listed in the appendix.

Commercial Products Here it is helpful to use Kamii and DeVries's criteria as a guide. Look for materials and products that *do* have options for variation. An example of a logical category of products is the marble rollway, of which there are many varieties. In some (which we would not recommend), the rollway is completely preset; all the child does is put the marble in the top and watch it roll down. In some, the angles of the rollway are predetermined; the child can only vary the height of the construction. Most of these structures are made of opaque plastic; there is one version, however, that we are particularly fond of. It is made of clear plastic so that children can watch the movement of the marbles more easily. In another type of marble rollway, the child can also vary the angle of incline. In still others, curved tracks can be produced.

Similarly, there are many commercial products that incorporate gears—with the same diversity of possibilities, depending on the product.

Which products are likely to be most useful to your program will be determined partly by the age, diversity, and experience of your children. Some of the rollways—for example, one with the *most* options—are much too difficult for even 5-year-olds to put together without a lot of adult assistance. These are more appropriate for children in, say, the second and third grades. And while a marble rollway with fixed inclines may not be optimally variable, it can provide lots of opportunities for other sorts of experimentation. This is particularly true if you have plenty of pieces available.

Other Materials It is essential to have many noncommercial materials on hand for developing and extending object movement activities. Some that we have found particularly useful are listed below:

> *Balls*—lots of them, of varying sizes, densities, surface materials (e.g., tennis balls versus smooth plastic balls), and weights (metal ball bearings and hollow plastic balls). Remember George Forman's principle of the continuum of change. It is useful to try to have a series of balls that form a continuum—so you would have two other types of balls in between the metal ball bearings and the hollow plastic balls that fall within that range of weight.
>
> It is also important to have sufficient *numbers* of such objects, so that children can, indeed, experiment freely with the dimensions involved.
>
> *Tubing*—particularly plastic tubing, but also tubes of all sorts, including cardboard tubes of varying lengths and diameters. Here parents can be very helpful, particularly around the holidays, with the cardboard tubing that usually holds wrapping paper being available. It is

also possible to locate community sources for tubes. Some factories regularly dispose of them; some copy centers also have paper tubes that they can save for you.

Pulleys, hooks, and ropes—since many activities that can be designed involve hanging objects and swinging things, it is useful to have on hand ways to safely suspend things from the ceiling. Your ability to do that will vary considerably from classroom to classroom. Even one big strong hook to which you can connect things is a help and usually can be installed without too much trouble. Better yet is a supporting wood beam in which you are free to drill holes and attach things. If these are not possible, a classroom can make do with one strong dowel suspended across the ceiling to which you can attach things.

Boxes—many movement activities can be developed and implemented with boxes of many different sizes and shapes, and you should get into the habit (if you are not already) of saving and scrounging them. We are always particularly pleased to obtain very large, long boxes that can be used for long inclines. Again, parents who are alerted to the importance of such seemingly disposable objects can be a good resource for you.

Extending Activities

Once an activity is developed or adapted, you have only begun! Each activity is really the seed of many more if you ask the right questions and observe the children who do the activity. A good activity will lead logically (or sometimes surprisingly) in many different directions, and the integration of the curriculum is greatly enhanced if you pay lots of attention to the outcome of the activity that you planned.

We will give one example and try to draw out some of the general principles on which you can extend *any* activity that you might develop or use.

■ The preschool teacher set up a simple maze with unit blocks on a table. A basket of table tennis balls and straws was placed beside the maze. The teacher expected the children to move the balls through the maze by blowing on them with the straws. One of the children, Alice, decides to hit the ball with a straw to make it move. Because the straw was a flexible one, it was not an effective "ball pusher." Setting the straw down, Alice goes to the easel area, takes a brush from the paint cup, and returns to the maze activity. She is able to move a ball through the maze quickly using the paintbrush. Pushing the ball along with the brush, she notices the trail that it leaves. She calls over to Amy, "Look at my path!"

Now let's think of some extensions. Rather than using blocks, create a maze with vertical pieces of rigid cardboard stuck together in such a way that you can lift up the cardboard to put fresh paper under it. Use paint-dipped balls to move through the maze, or have the children move paint-covered brushes through it.

Let's say that, after observing Alice, you planned a paintbrush-moving activity. In this activity, you observe one child following a path through the maze using one continuous brushstroke. After he lifts his painting out of the maze, he sets it on the table, leaves it, and then comes back in 5 minutes to see if it's dry. He traces his finger along the dry paint lines, which duplicate the maze pattern. From this observation, you might design a further extension in which children recreate the maze using their "maze maps" and unit blocks.

The first principle is, then, to *observe* children carefully. Often they have many ideas about how to extend the activity that are revealed through their play. This extension can either be done immediately, the result of quick thinking and response by the teacher, or it can be "filed away" for modifying the activity in a few days, even a week or two after more children have had a chance to interact with the activity as it is.

The second principle is for you, the teacher, to come up with a way to *challenge* what the child has already done. In a sense, you need to introduce an element to the activity that is puzzling or that even conflicts with what has been going on previously. This introduction of conflict or contradiction can make for some very interesting experimentation as children try out the ideas that "worked" before, only to find that they must modify them, given a new material or different objects. Let's look at another extension of our ball/maze activity that introduces some contradiction to an already familiar activity:

■ The teacher makes "crazy balls" by taking little plastic balls and, having poked holes in them, filling them with paint and putting a piece of tape over each hole. (When the children start to use the balls, they will take the tape off.) The teacher begins the activity introduction by asking the children what ways they have been using to move their marbles through their box-lid mazes. "Well," Judy answers, "you tilt it down the way you want it to roll, like this," using her hands to illustrate her explanation. "And if you want to move it a little bit over, you put it down this way," Billy interjects, using his hands also. The teacher holds up a crazy ball and explains, "I have these balls filled with paint. If you take the tape off here and roll the balls, any predictions about what might happen?" "They'll roll and paint will come out!" Judy answers quickly. "Do you have any predictions about how they'll move?" "Like a ball," Pam suggests. The teacher generates a few more predictions and then tries moving a ball with the tape removed. Its jerky movements as it changes weight from the release of the paint surprises the children. The teacher ends the introduction by saying, "This activity will be at the paint area where you can experiment with these 'crazy balls.' "

Molly picks up a ball, removes the tape, and puts the ball in the small box-lid tilt maze that had been used on previous days with marbles. She slowly tilts it and, as she does, sees the ball roll across the box. "Spots, it's making spots!" she says excitedly. She tilts the box lid farther, laughing as the "dotted lines" cover the cardboard. After all the paint has run out of one ball, she pulls the tape off another and puts it in the box next to the "empty" ball. Tilting the box gently, she notes that the balls move differently. "Hey, this one wins the race!" she calls.

Here the children who had already had previous experience tilting balls through mazes are challenged by a ball that moves in unpredictable ways. Extensions serve to add increasing complexity to an already familiar activity, or they can introduce a new, perhaps conflict-producing element to the activity.

Connecting the Curriculum

We want children's experiences across all areas of the curriculum to be as integrated as possible. Why? Because the more different ways the child connects ideas across media, across areas, and across activities the more *sense* those ideas make to children. We really know that a child understands an idea when, faced with a completely new situation, he or she applies lessons learned earlier in a different context.

How can we carry over to other areas of the curriculum some of the ways in which children experiment with movement? We can think of this in two ways: (1) directly connecting experiences by varying the scope and scale of the activities we present and (2) indirectly connecting experiences by creating an environment conducive to thinking about things in certain ways.

Direct Connections   One direct connection is with literacy activities. Reflecting with the class or with one child at a time about a movement experience and putting those reflections in writing can serve two important purposes. First, through reflection, the child's experience itself is recalled and enhanced. By verbalizing what he or she did and by hearing what others did, the child is, in effect, reenacting his or her activity. The second purpose lies in providing children with an exposure to print and to the meaningful uses of written language.

Let's look at how a teacher might conduct a written reflection with a group of children after an experience involving the movement of objects.

■ After activity time, the kindergarten children reconvene as a group before going outside. The teacher sits next to an easel covered with several layers of blank paper. "I noticed children making a lot of different marble ramps today. Since this activity will be set up all week, I thought we might make a list of problems children had when they were building their ramps. What should we call our list?" Connie volunteers, "How about 'Marble Moving Problems'?" "Okay," says the teacher, and as she writes she speaks the words slowly to correspond with her writing. "John," asks the teacher, "I saw you working very hard. What kind of problems did you run into?" "Well," says John, "I tried to make my ramp go around a corner and it fell down." "Oh, what should I write?" asks the teacher. "Write 'Ramps can fall down when you go around a corner, but you can prop it up with a block.' " The teacher writes down John's words and Linda pipes up, "You can also tape it to the floor so it won't fall." "That's another way to solve the problem. I'll write that too." After writing down more problems and ways to solve them, the teacher rereads the title and asks the children if there is anything they would like to add to the title. "How about 'Marble Moving Problems and Ways to Solve Them,' " suggests Connie.

This same process would be appropriate for older children. They also might enjoy mapping their constructions, labeling their maps, or even labeling their construction with directions.

Connecting with Books   Some books can directly enhance movement experiences. Just to give an example, the book by Ezra Jack Keats called *The Snowy Day* illustrates a stick leaving a track in the snow. This is a good example of the representation of motion, and it could be pointed out as the book is read to children who have engaged in an activity that involves leaving tracks of motion. Another example of a book that can directly enhance motion activities is Matt Novak's *Rolling*.

Now that we have presented some categories and some things to consider in designing activities and selecting materials, we shall present two movement activities as illustrations.

HOW CAN I MAKE IT MOVE? CONSTRUCTIVIST PHYSICS: SAMPLE ACTIVITIES

Purposes of Movement Activities

The following are general purposes that apply to most movement activities:

- to provide opportunities for action on objects that make the objects move
- to provide opportunities for experimentation with directing and/or representing movement
- to provide opportunities for experimenting with and constructing relationships between different variables
- to create an environment in which children ask questions, experiment, and build theories
- to facilitate children's mental representation about movement and to encourage their verbalizations about it

In order to better envision what a teacher might do to fulfill these purposes, we shall describe two movement activities in some depth.

FIRST ACTIVITY: TARGET PENDULUM

In this activity, children will be experimenting with the properties of a pendulum and directing its movement toward constructed targets.

Materials
- lightweight rope with a hook at the end
- three plastic shampoo bottles, one empty, one half-filled with sand, and the other filled with sand; the bottles will have to be taped shut securely and have a loop attached at the top
- approximately ten empty half-gallon cardboard milk cartons that have been sealed shut

Context and Brief Description of the Activity   In order to facilitate children in their roles as experimenters and scientists, we must create an environment of self-direction. In this movement activity, children will have the opportunity to experiment with knocking over milk cartons by swinging a pendulum that is hung from the ceiling. [*Note:* if you do not have a place on the ceiling where you can secure the rope, you could tie a rope from wall to wall and then hang the pendulum rope from that horizontal rope.] The children may change the differently weighted bobs at the end of the pendulum rope and rearrange their targets (the milk cartons). This activity might take place in one section of the room set up to accommodate a small number of children while the rest of the class is engaging in other projects. In this context, children would be free to explore and test their theories and then move on to different activities, allowing room for new children to participate.

Introducing the Activity   The following is a scenario describing one possible way this activity might be introduced to a group of kindergarten children. This particular introduction will focus the children's attention on the general goals of directing movement toward a target and encouraging both experimentation and predictions of the results. During the course of an introduction, children's comments or responses to the teacher's questions might focus attention on other goals.

For this introduction, you will have to be in the area of the room where the pendulum is set up or have another adult or child hold the rope to which the pendulum is attached. You should have the targets and the other bobs visible to the children. One way to introduce the activity is to ask the children how you should swing the pendulum so that it hits the milk cartons, or you could ask them to make predictions about what might happen if you were to swing the pendulum in a specific direction. Ask the children to tell you "using words" and then try out some of their ideas while following the children's verbal directions. For example, a child might tell you to swing it "real hard that way so that it goes high up in the air," predicting that it will hit the milk cartons. You repeat his or her predictions and then swing it "real hard" to see what happens. It is helpful to repeat the child's predictions in this manner so that all the children can hear and become involved in the testing process.

■ The kindergarteners are gathered in the area of the room where the swinging pendulum has been set up. The children are seated on the floor in a large circle around the teacher and the activity materials. Three cardboard milk cartons are stacked vertically about 4 feet from the bob. The teacher touches the pendulum to make it move back and forth slightly and points to the stacked cartons, saying, "I have this swinging pendulum. I wonder how I might swing it to hit these milk cartons and knock them down." Sarah quickly volunteers an idea, "I know . . . I know . . . swing the hanging thing around in a circle and it will hit them." The teacher repeats Sarah's words and says, "Sarah predicts the bob will hit the cartons. Are there any other predictions about what would happen if I swung this bob in a circle?" "It will keep going around and around after it hits the cartons," volunteers Kenny. The teacher follows Sarah's suggestion, but the bob just misses the cartons and then begins to swing in a circle. "What could I do differently?" asks the teacher. "How about swing it real hard that way," Carrie excitedly suggests as she points to the cartons. The teacher repeats her suggestion and swings the bob toward the cartons. The bob hits the carton this time but does not knock them down. Then she picks up a heavier bob, shakes it and says, "This one feels different. Let's try it." The teacher replaces the lightest bob with the heaviest one and swings it according to Carrie's suggestion. The cartons fall and one of the children shouts out, "Now try the other one," pointing to the middle-weighted one. The teacher suggests that he try that idea during activity time, and adds, "There are lots of different ways to swing the bobs and change them and to stack the targets. You can try out your ideas if you want to choose this activity during activity time."

How Children Might Engage in This Movement Activity   The following is a list of some experiments you might observe children doing with this movement activity. A child might spend the entire time experimenting with one possibility, or he or she might explore several possibilities. Those listed here are not intended to be sequential, nor are they a complete list of things children might do. Sometimes, one hopes, children will engage in an activity in ways we might not expect. For example, one time a child hooked all the bobs on at once and then swung the pendulum at all the milk cartons, which she had carefully taped together with masking tape.

- Children might swing the pendulum in different directions, as around and around or back and forth, without focusing on the targets.
- Children might apply varying degrees of force to the pendulum.
- Children might change the bobs and swing the pendulum in different directions with varying degrees of force.

- Children might push on the bob and watch the swing of the pendulum.
- Children might pull the bob toward themselves and then release it and watch the swing of the pendulum.
- Children might stack the cartons vertically or horizontally and change the distance of the milk cartons from the arc of the pendulum swing.
- Children might grab the rope of the pendulum right next to the bob and hit the target with the bob without releasing their hold on the rope.
- Children might talk about their strategies with each other, attempting to coordinate their efforts to knock the targets down.

Theories, Questions, and Predictions Children Might Have While Engaging in the Activity   As we discussed in Chapter 2, children do not necessarily verbalize the questions or theories that are underlying or "driving" their experimentation. However, if we could "get inside" a child's thought processes and translate them into our language, here are some things we might hear. (Remember, these are translated into *adult* language. Children would not verbalize their predictions in these ways!)

- How long will the pendulum keep swinging if I push it around just once? Will the circle of the swing keep getting smaller and smaller?
- If I push it real hard away from myself, what will happen? How long will it take for the pendulum to stop swinging entirely?
- If I push it softly, will the pendulum swing slower?
- How will the pendulum swing differently if I put on a heavier bob? . . . a lighter bob? If I push the heavier bob softly, will it move the same as when I push the lighter bob hard?
- What patterns in the movement of the pendulum can I see if I watch the bob move? Can I predict where it will move?
- How does the bob move if I *push* it, as opposed to if I pull it toward myself and release it?
- Where can I place the milk cartons so that the bob will hit them? Is it easier to hit them if they are stacked up tall or if they are in a long line? How does the weight of the bob affect the potential of the bob to knock down the target?
- What is the relationship between the weight of the bob, the force applied to it, and its effectiveness in knocking down the targets?

- Can I knock the target down by a more direct strategy, by holding onto the rope just above the bob and moving the bob without releasing it from my hold?
- What words can I use to explain to my classmates where to move the target while I make the pendulum swing?

This is but a sampling of the possible questions and theories the children might be experimenting with. It is important for you to observe the children and remain flexible enough to be able to facilitate any needed practical additions to their experiments. For example, a child might want to vary the length of the pendulum rope and experiment with the bob's distance from the floor at different points in the pendulum's swing.

Possible Extensions   Depending on the extent and focus of the children's initial experiments, you might want to provide for the following extensions of this activity. Extensions of an activity might occur within the initial activity period or over several days. Activity extensions might be based on the children's initiations, your observations of the children's experiments, and/or your projections of possibilities.

- The children might bring other objects to the project area to use as targets.
- The children might tape other objects to the bobs to vary the weight further and observe the results of this difference.
- The children might want to set up another pendulum alongside the already existing one and experiment with coordinating the swings.
- The children might want to create a game similar to tether ball, in which they hit the bob back and forth using paddles.

Two Scenarios

The first example is set in a preschool:

■ Andrew walks over to the swinging pendulum activity and gives the pendulum a light shove. It swings around, and after watching it, he places himself in the path of its trajectory so it will touch him. It gently bumps his stomach and, laughing, he places a milk carton where he has been standing and shoves the pendulum again. Noticing that the pendulum swings above the carton, Andrew adds another carton on top of the first one. This time when he swings the bob, he knocks over the top carton. "I did it!" he exclaims, as Franny comes over to observe. She picks up a heavier bob, shakes it, and says, "Let's try this one." The children experiment with changing the bobs, adding more cartons to the target and swinging the bob in different directions. They then decide to make a low long line of cartons stacked two high. "We're going to knock them all down!" Franny shouts. They swing the pendulum and it knocks one down. After several unsuccessful attempts to knock the others down, Andrew holds the rope just above the bob and knocks more cartons down without releasing his hold on the rope.

Now let's look at the same activity as it might transpire in a first-grade classroom:

■ Immediately after the activity introduction, Jan goes to the swinging pendulum activity and picks up the bobs, weighing each of them in the palm of her hand for a minute. She selects the lightest bob and, after placing it on the rope, swings it softly. Then she hits it progressively harder, watching where it swings. When Mark comes over to watch, he suggests, "Put the carton at the end of where it swings." Jan stacks three cartons together where Mark had suggested, steps back, pulls the bob toward her, and releases it. "We did it!" Jan and Mark both exclaim simultaneously, laughing. Next, Jan selects the lightest bob and directs Mark to place two cartons closer to her. She experiments with different amounts of force and then gives Mark directions for placement of the cartons. Mark suddenly says, "Hey, let's place *two* bobs on the hook together. Jan does that and the children continue to vary both the amount of force applied to the bob and the placement of the cartons.

SECOND ACTIVITY: MARBLE ROLLWAYS

In the second movement activity, designed from the question "How can I make it move?" the focus is more on creating ways to direct movement.

Materials Needed
- various kinds of cardboard tubes, including wrapping paper tubes, paper towel tubes, and toilet paper tubes; prepare the tubes by cutting most of them in half lengthwise, leaving a few tubes intact

- unit blocks of varying sizes
- marbles, at least two dozen
- masking tape

Context and Brief Description of Activity   In this activity, children will create rollways for the marbles using the lengthwise-cut cardboard tubes as chutes. Tape and blocks will be available to create these rollways of varying lengths, inclines, and configurations. Part of the floor area of the room out of the flow of traffic would be most appropriate. If this is not feasible, the activity may be conducted on tables shoved together.

Introducing the Activity   As we discussed in the introduction to the previous movement activity, the primary purpose of introducing an activity is to foster inquisitive attitudes and to encourage children's experiments. In order to plan a stimulating and appropriate activity introduction, you must look back at the goals and rationale for your activity. For example, in order to focus the children's attention on directing movement, you might want the children to consider the challenge of how to make the marble move around a corner. One way this might be accomplished is by first showing the children the cut tubes, blocks, and tape. You might pose the following question to them: "Does anyone have an idea about how I might move the marble around this chair using these tubes and tape?" Then actually demonstrate one or two of the different ideas the children offer, pausing before releasing the marble to ask if there are any predictions about where the marble might end up. Remember there is no one "correct" way to introduce an activity. The important thing to ask yourself when planning an activity introduction is "What are the goals that I want to focus on?" and to present the activity in a way that fosters the generation of ideas.

The following scenario illustrates some possible responses children might have to this teacher's introduction and questioning strategies.

■ A group of first-graders gathered around the teacher watch curiously as the teacher slowly lays out the tape, cut tubes, marbles, and blocks on the rug in front of the children. "Does anyone have some ideas about how I could make the marbles move in different ways using these materials?" asks the teacher. "You could tape some tubes together to make a long road," suggests Jackie. The teacher asks Jackie how many tubes to use and then proceeds to follow her suggestion, taping five tubes together, end to end. "Now how can we make the marbles move?" she asks the children. "How about putting the marbles at one end and hitting them with that skinny block?" offers Billy. "That's one way to do it. Are there any predictions how far it would roll if I hit it this hard?" asks the teacher, swinging the block to demonstrate. Jackie thinks it will roll all the way to the wall, but the other children agree that it will stop when it bumps into the edge of the rug. After hitting the marble and checking predictions, the teacher asks the children if they have some ideas about how they could make the marbles move without hitting them. The children offer their ideas—such as lifting up the taped-together tubes and taping one end to the edge of a chair, elevating the tubes with blocks to make them go up and down, and cutting little pieces of tube and taping them together to make a curved road. The teacher ends the activity introduction by saying, "I will be interested in seeing all the different structures you can build to make the marbles move."

How Children Might Engage in This Movement Activity   Children might work individually or collaborate in any of the following ways:

- They might make a series of individual inclines of varying degrees and heights.
- They might build a continuous ramp with a series of peaks.
- They might build a marble rollway that curves or changes direction on a horizontal plane.
- They might try to see how steep a ramp they can build.
- They might see how long a ramp they can build.
- They might build a rollway that is elevated off the floor.
- They might build a ramp, set up a target at the end of the ramp, and roll marbles down the ramp at the target.
- They might send different-sized marbles down their rollways.

Theories, Questions, and Predictions Children Might Have While Engaging in the Activity   Again, remember these are not *real* questions the children ask but are stated in adult language.

- How does the degree of the incline affect the speed of the marble?
- How does the degree of the incline affect the distance the marble will roll?

- What is the relationship between the height of the incline and the degree of the incline, given the same length of cut tube?
- Which rollway will cause the marble to roll further—a steep, short one or a long, gradual incline?
- What degree of incline will cause the marbles to roll up and down peaks?
- What angle will allow the marbles to roll around corners?
- How can I balance my structure so that it is elevated off the floor and the movement of the marbles rolling will not tip it?
- How will the degree of the incline affect the force at which the marbles will hit the target? Does this force diminish with distance?
- How does the size of the marble affect its movement?

Possible Extensions

- Children might request more tubes and collaborate to build a larger rollway that extends beyond the activity area and around the perimeter of the room.
- Children might make rollways at very steep inclines and place a block at the bottom, attempting to make the marble bounce up into a basket.
- Pulleys and string could be supplied for the children to create a pulley system to vary the incline of the rollway.
- A similar activity on a larger scale using balls and large tubes and chairs could be set up in a large area of the room or outside.

A Possible Scenario

■ Six-year-old Judy eagerly approaches the corner of the first-grade classroom where the teacher has set out cardboard tubes cut in half lengthwise, a container of marbles, a pile of unit blocks, and masking tape. Without hesitating, she lays out some tubes on the floor, compares their length, and announces, "I'm going to make the marble move up and down and around!" She places one of the longer unit blocks vertically and leans a tube against it at a very steep incline. As she reaches around to pick up another tube, she bumps the block and propped-up tube over. Changing her strategy, Judy places the block horizontally and props one tube against the block in one direction and another in the other direction to form an "up and down" pathway for the marble. Sarah tapes the tubes together where they meet and then proceeds to make two more, similar structures, joining them together. As she places a marble at the top of one of the peaks, she says, "This is going to roll down this side to the box!" When she releases the marble and it rolls down the opposed incline, she looks surprised and begins experimenting with releasing a marble from the peak of each part of her construction. After 10 minutes she calls to Ted, who is reading a book nearby, and says, "Here, you hold a marble and I'll hold a marble and we'll let go of them at the top of my mountains and see if they meet." Ted puts his book down and the two children begin Judy's "game," laughing every time the marbles run into each other.

The scenarios presented in this sample activities section are but a few ways that children might experiment with movement. The constructivist teacher needs to plan and prepare activities based on carefully thought out expectations yet be flexible enough to support different ways children might ask the question "How can I make it move?"

Chapter
6

How Can I Make It Change? Constructivist Chemistry

What comes to your mind when you think of chemistry? Probably periodic tables, beakers, bunsen burners, and funny smells. These are the tools that grownups use to understand the structure and composition of matter. If you think about what you are doing as you heat a funny-smelling substance as you add that white powder to the brown one, or as you calculate the resulting molecular structure, you'll see that all you are doing is transforming substances—changing their composition in order to understand them. Since transformation is such a natural focus of young children, we'll use it in defining early childhood chemistry.

EARLY CHILDHOOD CHEMISTRY AS TRANSFORMATION

Transformation—the word captures much of what children are engaged in and focus on as they play. Play dough, pushed and pounded, changes from a ball into a pancake; blocks, one on top of another, change from a wall into a tower; a cardboard box painted in various colors changes into a playhouse; all these activities involve transformation. Some of the transformation that children engage in is physical; that is, they act to change the shape, form, or substance of materials. Some of the transformation is symbolic; children imagine the changes they want as objects "become" other things or as the child "becomes" another person. In this chapter, we'll describe some ways to develop activities and present materials to focus on transformation, viewing transformation as an integral part of scientific understanding. Some of the activities may not seem like chemistry. But if you take the essence of chemistry and think about the ways *children* are capable of understanding science, as we have been encouraging you to do in previous chapters, you will start to see how focusing on the transformation of objects and materials is comprehensible to young children. It makes sense to them and can later lead them to an interest in and understanding of the grownups' world of chemistry.

How This Approach to Chemistry Is Different from Other Approaches

If you look at books and activities designed to present chemistry to young children, you will notice first that most are geared only to older children. There is something about chemistry in particular that is seen as not appropriate for the younger age group we are considering here. Maybe it's because much of chemistry is "invisible" or microscopic, or because much of it involves potentially toxic or dangerous substances and materials.

And even when you see "chemistry for children," you will notice a real difference between most other approaches and ours.

One characteristic of many of these approaches is to focus on the mystery and magic of processes, particularly in the area of chemistry. The idea, presumably, is that by having them combine materials that produce unusual, amazing reactions, children will become intrigued, fascinated, and interested. Let's give an example of a nonconstructivist activity that might be in such a curriculum.

■ In a second-grade classroom, Billy and Whitney stand together at the table labeled "Solids," carefully scooping one spoonful each of baking soda, unflavored gelatin, and alum into their cups. Then they move to the table marked "Liquids" and each pours 3 spoonfuls of vinegar into a cup and adds a drop of food coloring. They walk back to their desks together. Billy whispers to Whitney, "What's next?" Whitney replies, "Listen for directions!" The teacher tells all the children to count to three and then all together pour their "solids" into their "liquids." "One . . . two . . .

three!" The children all pour and squeals of delight are heard around the room as the cups begin to overflow with foam.

What are the children learning as they interact with these materials? Is there a connection between their actions and the resulting reactions of the materials? Is there anything they can *do* to better understand what is going on? Are there ways they can vary their actions to produce different responses?

We are certainly not opposed to providing children with experiences that they cannot completely understand—indeed, that might even be mystifying. As we have discussed in reference to the important role of conflict and contradiction, the experience of the "sense of puzzlement" is an important element in the construction of knowledge. But for puzzlement to lead to the construction of knowledge, children have to be able to contribute to the interaction; they have to have *some* ideas, even if incomplete, about the phenomena they are observing. We would question what children really learn from experiences that mystify (as opposed to puzzle) them. Further, we would argue that such experiences cannot be the foundation or center of a science curriculum.

Why not? What *is* the role of "magic" and "mystery"? Let's think about this only in terms of transformation for now. Transformation is, simply, a change from point A to point B. In describing transformation from a constructivist perspective, we are interested in transformation of two kinds—physical and symbolic. Physical transformations occur when children's actions produce the change from point A to point B; symbolic transformations occur when children's imaginations produce the change from point A to point B. Both physical and symbolic transformations have their sources in the child—the child's own physical or mental actions.

"Magical" transformations, on the other hand, have as their source some other unidentified, unknown, or even unknowable process. They are puzzling—beyond the children's understanding. Yet, as constructivists, we want children to experience themselves as sources of knowledge and not to see knowledge as coming from an external source.

Of course, it may be that such magical transformations could lead a child to experiment further. In this case, the "magic" is only a lead-in to the experiment in which the child *is* the source of knowledge. Too often, however, the magical transformation is the whole point of the activity. Think about the "magic" focus involved in growing "magic rocks" from unknown chemicals or in mixing two chemicals together in a bottle with a balloon on top and watching the balloon blow up. Unless they were put in the right context, such activities would not lead to further experimentation and would actually just mystify the child.

These children are experimenting with one type of transformation—different forms of the same materials—as they mix shaving cream, sand, and paint together.

Categories of Chemistry Activities

There are two distinct categories that we are considering as chemistry. Both involve change or transformation, but in different ways. Some change activities focus on different forms of the same material, while others provide opportunities to experiment with rearranging the same recognizable parts. Unlike the categories described in Chapter 5, these two types of change activities are not sequential in any way; that is, one category is not more complex or advanced than the other.

The categories that we will describe involve (1) the rearrangement of parts and structures (reconstruction) and (2) changes in substance and consistency (combination).

Reconstruction Reconstruction involves the rearrangement of existing parts and structures to produce a new structure. Reconstruction can involve building and rebuilding with the same parts, or it can involve construction by adding new parts.

An important element of reconstruction activities is the child's ability to reverse his or her actions on the materials. Is it possible to go from point A to point B and then back to point A? Or is the reconstruction irreversible? While it is not necessary for all children's experiences to be characterized by the ability to perform reversible actions, part of the focus of at least some activities should incorporate reversibility. Why? Because reversibility focuses the child's attention on the process of transformation itself—not just on the product of transformation.

Let's consider some examples of reconstruction activities. We'll start with two examples of reversible reconstruction. Here is the first:

> ■ In a preschool, Diana grabs a big hunk of play dough from the pile and pounds it with her fist until there is a big hole in it. She looks at it, pokes her finger in the center of the hole, and tears off two pieces from the thickest part of her play dough. She pounds these pieces until they are flat and then begins poking holes in them, singing as she pokes, "The holes are coming in, the holes are coming in." Then Diana clumps up each flat piece and puts them all together, rolling the mass around with her palms until she picks it up and sets it back on the big pile of play dough.

And here is the second:

> ■ In a first-grade classroom, various "reconstruction" materials are placed on the shelf. In one area of the room, Cindy sits down on the rug and reaches for the Mobilos,® which have hinged, interlocking parts. She calls to Matthew, "Hey, let's make transformers." Both children immediately begin putting pieces together. Cindy puts together various pieces, swivels the hinged parts around, and says, "My car is changing into a helicopter." Matthew looks closely at what she has put together, looks at his "airplane," and takes it apart carefully. "I'm making a 'copter transformer too," he announces and starts to duplicate Cindy's construction.

In these two scenarios, the children could, if they chose, transform a material and then transform it back to its original state. Now we will give two examples of irreversible constructions, in which the children cannot reverse their changes. Here is the first:

> ■ In a preschool, Jessye dips a small wood scrap into glue and places it on a small piece of Masonite. She continues this process until the Masonite is covered with wood scraps. Pausing to look at what she has done, Jessye dips another wood scrap in the glue and places it on top of the glued pieces. She dips another scrap

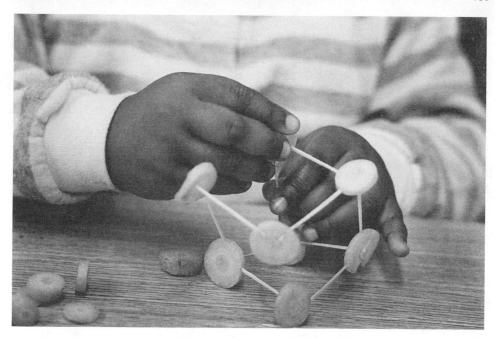

Rearrangement of the same recognizable parts is another type of transformation.

in the glue and announces, "Mine's getting bigger and bigger." Then she balances it on top of the wood piece she just glued.

And here is the second:

■ In one corner of a second-grade classroom the teacher has placed hooks at various levels and suspended loops of string from thc cciling. Extra chairs are placed on one side of the area for the room weaving project. Gary takes one end of the string and ties it to one of the hooks at eye level. He unrolls some string, looks up to the highest hook, and cuts the string. Next, he pulls a chair close to the highest hook and attaches the string to the hook. "This can be our starting place," he says to Mark, who is inspecting the placement of the hooks around the area. Mark cuts five strings of equal length and ties them to the horizontal string. Mark and Gary then both begin tying each of the other ends of the five strings to other hooks. Other children come to the room weaving area and contribute to its progress by attaching strings either to hooks and chairs and across the area or to strings already attached.

Combinations

The other category of transformation activities involves changes in substance or consistency, usually by combining materials. Many common and popular activities for young children involve combinations—of sand, dirt, water and paint. Children seem "naturally inclined" to pour, mix, add, stir, and especially stick their hands into gooey, wet messes. The focus in such transformation activities is the change from one consistency or substance to another: from powder to watery paint, from dry dirt to goopy mud, from white shaving cream to colored shaving cream.

This brings up the issue of "color mixing" activities. Color mixing is a common activity that might be interpreted as falling into this category. From our perspective, color is *one* attribute that may change as children combine materials. In that case, our focus would not be specifically on the color change that results; rather, it would be on *the process of change through combinations*. The goal of the activity would not be to teach the children, for example, that blue and yellow makes green, but that when two substances combine, one thing that may change is the color (or shade) of the substance. And the "continuum" of change would be more of a focus than the end points.

In generating ideas for activities that involve combinations, keep in mind the earlier discussion of magic and mystery. It is certainly interesting in some combination activities to get unusual results—that is, have the product of one or more combinations be of an unusual consistency, result in unusual products, or involve combinations that children don't usually get to initiate. But to *mystify* children is not the point.

Let's see how this might work.

■ After a week of providing flour, water, salt, and cornmeal for the children to experiment with, a kindergarten teacher set out cornstarch and water with bowls and spoons. Some of the predictions generated by the children during the activity introduction were that it would get thick and then, like flour, would get "really sticky." As Kirsten experiments with adding alternately water and cornstarch to her bowl, she exclaims, "Hey, this is weird! It gets real hard." "No it doesn't, look at mine," Greg says as he picks up a clump and watches it "melt" through his hands. Some drops on the table and Kirsten rubs it around with her fingers, lifts her finger up, and rubs it on her smock.

In this example, the goal of providing an unusual combination is not to mystify children. Instead, the unusual properties of cornstarch and water as it changes from liquid to solid can focus the children's attention on the fact that the same material can take different forms.

Here are some examples of other combination activities that focus on transformation.

■ In a first-grade classroom, Joe and Linda are seated at the "paint-making" table. Before them are containers and spoons of various sizes, water, powdered tempera, and flour. First Linda scoops out two spoonfuls of tempera and puts it in her cup. Then she fills the cup three-quarters full of water and stirs it. "Mine is making paint," she explains to Joe. "Mine's clay!" says Joe, whose thick mixture is predominantly tempera with a small amount of water. Linda looks carefully at Joe's, pokes it with a stirring stick, and then begins to swirl her "paint" rapidly. "I need to change mine," Linda informs Joe, as she adds some flour. She stirs it around, lifting the spoon out of the mixture occasionally and letting the mixture drop back into the container. She adds more water and then more flour alternately as she continues to inspect the results.

■ In a preschool, 4-year old Casey cautiously sticks one finger in the shaving cream "glopped" directly on the table. She smiles and says to Mandy, "It's puffy," "Yeah it's puffy," Mandy repeats. "Let's smooth it." Both girls begin patting the piles of shaving cream flat with the palms of their hands. Libby, who has been watching, notices the shakers full of powdered tempera. She picks one up and shakes it into the shaving cream. "I'm changing it pretty," she tells Mandy and Casey. The powdered paint makes depressions in the shaving cream and the girls begin poking the depressions and swirling the two substances together with their fingers.

Issues in Transformation Activities

Making Versus Observing Transformation   You may notice that the activities described above have one thing in common: they all directly involve children themselves in transforming materials. Children are changing things and observing the results of their own actions. Recalling the criteria for "good" physical knowledge activities described in Chapter 5, we may note that these activities, because of the child's own production of change, go far toward meeting those criteria.

There are, however, many ways in which children can observe transformation even though they cannot play a role in creating it. Such observations, while perhaps not very "good" activities, may still contribute to children's interest in and awareness of transformation in the world around them. This interest and awareness can contribute importantly to children's sensitivity to their environment.

Clearly, many of the events in the natural world involve transformation. Growth, decomposition, and the creation of habitats all involve transformations that children are not involved in creating but that can be observed. Because of this focus on the natural world, we shall be discussing observable transformation in greater depth in the next chapter. There our educational activities will be more limited in terms of the physical knowledge criteria that are essential to both movement and transformational activities.

Process Versus Product   It can be tempting, in change activities, to place too much emphasis on the *product* of the change rather than the *process* of change itself. The *changing* should always be the emphasis of the activity; this is not a "goal-directed" category. Change activities in which numerous "end points" can be achieved are great, because then there is a de-emphasis of the *one* end point. Here is an example of a change activity that focuses on continuous change.

■ In this change activity, a large block of ice is sitting in a plastic tub. Hammers, goggles, and shakers of salt are set nearby for the children to use. Marianne puts on a pair of goggles and pounds one corner of the ice block until a piece breaks off. She sets the ice chip on top of the block, shakes salt on it, rubs it around, and begins pounding some more. When she finally succeeds in breaking the chip into tiny pieces, she scrapes it off the big block. Next she pounds the center of the block with the hammer, stopping every so often to feel her results. She fills the indentation with salt, pounds it for awhile, and then rubs her finger on it.

The next day the teacher sets out the same activity with a fresh chunk of ice. This time the teacher elevates the tub, pokes a hole in the bottom of it, and inserts one end of a plastic tube in the tub and the other end in a clear, graduated container. The teacher hopes that this alteration will direct children's attention to the water from the melted ice, which drains into the container. As she has anticipated, the children periodically interrupt their chipping away at the ice to check the level of water in the graduated container.

In order to focus on the change process too, you would not want to overemphasize the "before" and "after" states, although it would be helpful for children to have access to materials in a variety of conditions. Think in terms of encouraging children to focus on the many states of the substance as it goes from point A to point B rather than just on A and B.

■ The teacher notices the preschoolers using their small squares of Masonite to flatten balls of play dough, so as to make pancakes. They place a clump of play dough on the table, put the Masonite on top of it, and pound. Each time they lift up the Masonite, they squeal, "I made a pancake!" The following day the teacher sets out play dough and, instead of Masonite, small flat pieces of clear hard plastic. The children again begin the "pancake-making" process. Now that the transformation is visible, the children can be heard saying, "My pancake's getting flatter and flatter" and commenting on the way the play dough changes.

Some Practical Constraints

Many transformation activities require two things that are often problematic at least at first glance. First, many require space. It requires commitment of space, often for extended periods of time, to be able to construct with giant building bricks or maintain part of the room for an ongoing weaving project.

If this is a problem, change the scale of the activity. Try a tabletop weaving project, use the smaller building materials, limit the number of children who can participate at one time in a smaller area of the room. If necessary, just select those ideas that fit your space.

The second problem is messiness, both of the physical environment and of the children. This requires some evaluation of your values and some creative problem solving on your part. Messes can often, at least in part, be minimized by the way the activity is set up. For example, if you are concerned about getting the rug wet, lay down large sheets of linoleum and use duct tape to fasten them. For smaller-scale wet messes, use large sheets of vinyl, available at fabric stores, which will wash off. For children, you must provide smocks or aprons. You must also warn parents ahead of time about the possibility of messes. In some classrooms, it is just a way of life that children *may* get wet or dirty; you can let parents know that they have to provide changes of clothes and that children should not come to school wearing clothes that cannot get messy. If you set up those expectations from the beginning, you will be teaching parents about the importance of "messes" in experimentation.

SOURCES OF IDEAS

Art and Household Supplies

Art and household supplies are excellent resources both for materials to be transformed and equipment to highlight the transformation process. In selecting "ingredients," consider how the properties of the materials might contribute to the child's theory building and whether some materials merely add a "magic" effect without contributing to conflict and contradiction. The more of these materials you can have on hand, the more scope you will have in designing new activities and facilitating children's activity extensions.

Materials that lend themselves readily to "changes in substance" include powdered and liquid tempera, flour, salt, cornstarch, shaving cream, soap flakes, sand, dirt, food coloring, and glitter.

All forms of clay and dough, including types that harden, lend themselves well to change activities focusing either on change in form or change in substance. All types and colors of paper should be available either to be used as paper or in combination with other materials such as water, which will alter paper's form. All types of tape, glue, and string should also be easily accessible to encourage children to create combinations and constructions.

There are several nonexpendable pieces of equipment that can be used for a variety of activities and help to focus children on the continuum of change. Whenever possible, use equipment that is transparent, so results of the child's interactions with the materials are visible. Clear containers of various sizes may be used for mixing and measuring. The mixing may be done in or over rubber tubs or a small children's swimming pool (to help control messes). Sieves, funnels, and clear flexible tubing can focus attention on the changing properties of materials. Stirring implements, paintbrushes, paint rollers, and eyedroppers may be used to facilitate the combination of materials.

A large, flat piece of clear hard plastic is an excellent resource. It may be used either vertically or horizontally to mix substances on, so that children may see changes from various sides of the mixing surface. A large piece of clear plastic may be framed with wood for durability and safety. Smaller pieces more easily handled by a child are also useful; their edges can be filed to make them safe.

Commercial Products

We have found the following commercial materials particularly well suited to transformation activities involving the rearrangement of existing parts. Ease in putting them together and taking them apart as well as the nonrepresentational nature of the materials make them particularly appropriate. If possible, supply materials with identical characteristics but in different-size scales—for example, regular-size and giant Tinkertoys®, Legos®, Duplos®, and large-scale interlocking blocks.

Mobilos®—this excellent construction material is made up of interlocking pieces with attachable hinges, so that the created piece is instantly transformable.

Duplos®, Legos®, and Large-Scale Versions—all are based on the same interlocking bricklike design.

Flexi-blocks®—these are similar to Legos® except that the pieces are hinged where they attach.

Tinkertoys® (Small and Giant)—small sizes are usually frustrating for preschoolers.

Googleplex® and Ramagon®—interlocking construction with integrated flat plates that are complex and appropriate for children in the primary grades, but too difficult for preschoolers to use easily.

Popoids®—bendable construction toy; best for transformation activities are the ones without representational elements. (Some of the sets are sold as "animals" and "trains.")

Extending Change Activities

Using Observation to Extend Activities   The child's theory building is the driving force behind curriculum in the constructivist classroom. In order to develop extensions of activities that reflect the children's form of experimentation, the teacher must observe carefully and understand how the children are interacting with the materials. This means the teacher must probe beyond what is seen at first glance. For example, consider the scene in which two children are playing "cook," mixing mud and water together and pouring the mixture into pie pans balanced on the edge of the sand table. A closer look and analysis reveal that what the children are actually exploring is the relationship between where they pour the mixture in the pan and how well it balances without tipping. The teacher then might plan an activity extension in which a 2- by 4-foot wooden board is placed down the center of the sand table, funnels are suspended over the board, and pitchers of water and pie pans are provided. Another thing that the children might do during this "cooking" play is to add some solid materials such as marbles to the mud and water, mix them, and then carefully tip the pan to see what stays in and what runs out. The teacher could presume that the children are experimenting with the properties of liquids and solids or perhaps with the ability of a solid to "thicken" a liquid. On the basis of these presumptions, the teacher might then plan an activity with sieves and materials that represent the continuum from liquid to solid. Or she might plan an activity with flour to be used as a thickening agent with mud and water.

By observing children carefully and thinking about the goals of a constructivist approach to chemistry—as well as the different categories of transformation—we can discover an abundance of activity extensions.

Introducing Conflict and Contradiction

Sometimes a teacher might choose to introduce some conflict and contradiction into the children's experimentation. As we discussed at the beginning of this chapter, a science curriculum cannot be based upon "magic" activities in which the properties of change remain a complete mystery. However, sometimes a "strange" activity introduced at the proper time can cause children to question or solidify their theories. For example, we have described a scenario of two children experimenting with cornstarch and water. Cornstarch has unusual properties when mixed with water—it changes from a solid to a liquid merely by being handled. A teacher might introduce cornstarch after the children had been exploring the properties of flour and water, experimenting with the way the mixture changes and becomes more like a liquid with the addition of water and thicker with the addition of flour. Perhaps the introduction of cornstarch, with its unusual properties, would focus the children's attention even more on the changes from liquid to solid and back again.

Another example of introducing "strangeness" to an activity would be the addition of salt to ice after the children had been experimenting with chipping it and observing how the various sizes of ice chips melt.

The key in introducing conflict and contradiction is to follow up on your observations of children. Is the "strange" element contributing to the experimentation? Or is it too "mysterious" and thus mystifying children? Does it promote theory building? These questions should guide all your activity extensions, but they are particularly critical here.

Across the Curriculum

The focus on transformation can be meaningfully extended across all curriculum areas. Here we'll give just a few examples to illustrate how a theme that is based on children's thinking about the world can meaningfully extend to many aspects of the classroom experience.

Many books focus on transformation. Two of our favorites, again just as examples, are *Changes, Changes,* by Pat Hoban and *Round Trip* by Ann

Jonas. Both these books do a remarkable job of illustrating and carrying children through the processes of physical transformation and back again, highlighting reversibility. Many folk stories or fantasy books, too, have transformation as a theme. Think of *Sylvester and the Magic Pebble* by William Steig or *Alice in Wonderland* by Lewis Carroll as examples.

Other possible cross-curriculum extensions include the incorporation of computer graphics programs that can "transform" drawings (changing shape, magnifying child-created drawings, mirroring images, etc.) and a focus on the transformations of the natural world, which will be discussed in the next chapter. Field trips to factories or other "production places"—where children can observe the transformation of objects from one form to another (e.g., a wood mill or an assembly line)—can also sharpen their awareness of transformation.

Finally, one of the richest possibilities of extension lies in the area of dramatic play, for, as we said at the beginning of this chapter, the child's world is *full* of transformation, and symbolic play is an important part of the way children constantly change the world around them. Permitting and facilitating symbolic play is an essential element of a classroom that encourages transformational thinking.

HOW CAN I MAKE IT CHANGE?
CONSTRUCTIVIST CHEMISTRY: SAMPLE ACTIVITIES

Purposes of Change Activities

As we described previously, some change activities focus the child's experimentation on different forms of the same material, while others give the child opportunities to experiment with rearranging the same recognizable parts. The first sample activity focuses on the changing of form. The second activity, presented more briefly, focuses on the rearrangement of parts.

The general rationales for activities that focus on change are as follows:

- to facilitate children's awareness that objects can change substantially in form and/or structure
- to facilitate children's awareness that change can occur along a continuum, gradually, rather than suddenly, in a leap from one state to another state
- to facilitate children's awareness that some changes are reversible
- to encourage children to be flexible in the ways they perceive the world around them
- to facilitate children's mental representations of change and encourage their verbalizations about it

The form of toilet tissue changes substantially when it is mixed with water.

- to create an environment in which children ask questions, experiment, and build theories

FIRST ACTIVITY: "PAPER MUSH"

In addition to serving as an illustration of an activity that focuses on changing form, this first activity illustrates how expensive equipment and extensive preparation by the teacher are not necessary. In fact, by selecting everyday materials, we give children the opportunity to experiment with familiar objects in unusual ways, encouraging perspective taking and flexible, creative thinking.

Context and Brief Description of Activity   In order to facilitate children's roles as experimenters and "scientists," we must create an environment of self-direction. In this particular change activity, children will have the opportunity to experiment with varying combinations of water and toilet paper. There are several ways to present this activity which satisfy the requirements of self-direction and free exploration. One possibility is to have one section of the classroom set up to accommodate a small number of children engaging in this activity while the rest of the class is occupied with other projects. In this context, children would be free to explore and test their theories and then move on to other activities, making way for new children to participate. Another possibility, depending on the number of students and the supplies available, might be to have the entire class engaged in this activity, perhaps in groups of three or four children, with other choices open to those who finish experimenting. Whatever the context of presentation, the teacher must assume the role of facilitator. As such, the teacher might step back from the activity and observe the ways in which the children are exploring. Such observation might yield information on possible ways to extend the activity or on the types of questions the teacher might ask.

Materials   Note that the quantities of materials will vary depending on how many children will be engaged in the activity at any one time and whether they will be working individually or in spontaneous or planned small groups.

- toilet paper, approximately a quarter to a half roll per child.
- large bowls or plastic tubs
- pitchers
- water
- a table or tables in an area where the floor is protected
- plastic smocks or old shirts to protect clothing
- food coloring
- glitter (optional)

Ways to Introduce the Activity   Before the children actually begin their self-directed exploration, you should "introduce" the activity either to the entire class or to a smaller group of children. Be sure to have water and toilet paper visible while you make the introduction. One possible way to introduce the activity is to ask the children if they have some ideas on how to change the paper. Or you might ask them to make predictions about how toilet paper might change if you added it to water. Ask children to "tell you using words"; then try out a few of their predictions, following the children's verbal directions. For example, a child might tell you that the paper

---

We have found that children respond positively to discussions regarding the special nature of using toilet paper in this context and understand when it might not be appropriate to use. We feel that toilet paper's readily visible changing properties make it worthy of use. However, if you object to using this material, other types of paper—such as tissue paper, napkins, paper towels, or blank newsprint paper—can be substituted.

will get mushy in the water and direct you to unroll a designated amount of paper, put it in the water, stir it around, and then lift it out of the water. Sometimes it helps if you reiterate the child's predictions so all the children can hear and become involved in the hypothesis-testing process. End your introduction with some concluding statement that emphasizes the range of possibilities, such as, "For those of you that are interested in experimenting with changing this paper, this activity will be on the round table," or, "These are some predictions. I'm interested in all the different ideas children have about changing the paper."

The following scenario shows how such an introduction might go:

■ The teacher is seated on the floor with a group of fifteen 4- and 5-year-olds. In front of him is a clear pitcher of water, a large clear bowl, and a roll of toilet paper. He assumes an exaggerated, puzzled expression, holds up the roll of toilet paper, and asks, "I wonder how I could change this toilet paper?" Peter eagerly responds, "I know, I know, I have an idea! You could pour water on it!" The teacher repeats Peter's idea by saying, "I could change the toilet paper by pouring water on it?" and pretends to begin to pour water on the toilet paper. "No, no," says Peter, laughing, "first you tear off some pieces and put them in the bowl and then you pour water on it." The teacher follows Peter's suggestion and holds the bowl for all the children to see. He asks, "Are there any predictions about how the paper would change if I lifted it out of the water? Adrienne answers, "It would get mushy in your hand." "Any other predictions?" asks the teacher before lifting the paper out of the water. "It would break?" questions Amy. The teacher lifts the wet toilet paper out of the water and holds it for the children to see. After watching the part of the paper that hangs down, breaks off, and plops into the bowl, Amy directs the teacher, "Squoosh it together . . . squoosh it in a ball." After eliciting a few more suggestions from the children, the teacher concludes, "You have a lot of ideas about changing this paper. This activity will be at the back table for those of you who are interested in experimenting with your ideas."

How Children Might Engage in This Activity   The following is a list of some possible experiments you might observe children engaging in with this activity. Once you create an environment that encourages experimentation, there will be a broad range of possible ways children might engage in it. Remember, children will come up with experiments you could not have predicted. For example, one time a child carefully unrolled some toilet paper around the table and room to "measure" the length of the paper and the floor surface before she submerged the paper in the water!

- Children might unroll large sections of the toilet paper, clump them together, and submerge them in the water.
- Children might pull off one section at a time, put the sections in the water, and then observe and feel that the toilet paper has become mushy.
- Children might put the entire roll of toilet paper in the bowl, pour water over it, squeeze the roll, and feel the water come out of it. Or they might pull pieces of the wet paper apart and then add more water.
- Children might add color and glitter to the water before adding the toilet paper. As the toilet paper breaks apart, the color and glitter will become more and more mixed with it.
- Children might add food coloring to the water and toilet paper, stir it around, and then squeeze the toilet paper together into a ball. The color will fade from the toilet paper as the moisture leaves it.
- Children might put toilet paper in the bowl and add water a little bit at a time.
- Children might take a wet, whole roll of toilet paper out of the water, set it on the table, and pound it flat.
- Children might squeeze separated, mushy bits of toilet paper together into a ball and set it out on the table to dry, or they might place it into the water again and break it apart.

Theories, Questions, and Predictions Children Might Have While Engaging in the Activity   The following are "adult translations" of some of the questions that may be driving children's experiments in this activity. Remember, children do not necessarily verbalize or even think these questions. But if we could interpret a child's thought processes, here are some of the theories, questions, and predictions we might find.

- What if I unroll the toilet paper and put it in the water? How much space will it occupy?
- What form will the toilet paper take when I put it in water? How will it change? How quickly will it fall apart?
- How can I speed up the change process? What will happen if I pull the paper apart while it is in the water?
- What will happen if I pull off a section at a time and place it in the water? How can I make it "dissolve" all by itself? How can I pick it up when it is no longer in the same form?

- What will happen if I put the whole roll of toilet paper in the bowl and then pour water all over it? If the water goes into the roll of paper, how can I get the water out of the paper and back into the bowl?
- What can I do to make it easier to pull the roll apart? How will water change the ease with which the paper pulls apart?
- What will happen if I add food coloring and glitter to the water with the broken-apart toilet paper in it? Where will the color and glitter go if I squeeze the paper into a ball?
- How can I change the toilet paper back into something flat?
- What will happen if I shape it into a ball and set it out to dry?

These are but a few of the *possible* experiments children might engage in. As a teacher, it is essential for you to be attentive and flexible in order to facilitate the unplanned additions to the children's experiments. For example, a child might ask for a slotted spoon so as to experiment with the changed state of the paper, exploring whether or not the wet paper can flow through a slotted spoon or through sieves of varying sizes.

Possible Extensions   The following are some possible ways in which children might naturally extend this activity. Even if they do not come up with these extensions, you might want to plan activities based on these ideas.

- The children might clump together the broken-apart toilet paper into some form and observe how it changes as it dries.
- The children might reverse the change process after the paper has dried by wetting it again.
- The children might try different types of paper and other materials and experiment with the way they change in water.
- The children might experiment with adding other materials (glue, for example) to water and paper to see in what ways they stay mixed with the water and in what ways they cling to the paper.

Two Scenarios

The following two scenarios illustrate children's experimentation and theory building in the paper-changing activity. Here is the first:

■ Tommy, age 3, goes to the table and begins unrolling the toilet paper into a large pile without using the water in any way. His eyes focus on the roll as it decreases in dimension; he is unaware that the large unrolled pile has begun to sink into the water. The observing teacher, aware of Tommy's focus, does not point this out to him. After a while, Tommy turns to the sinking pile of unwound toilet paper and looks puzzled. He then plunges the pile of unrolled toilet paper into the water with both hands and begins squeezing it in the water, lifting it out of the water, and squeezing it again. When Tommy stops squeezing the toilet paper, the teacher asks, "I wonder what would happen to the toilet paper if we added coloring to the water?" Tommy, not responding directly to the question, replies, "I want green." The teacher adds green coloring and Tommy swirls it in the water and begins again to squeeze the toilet paper in the water and then lift it out of the water to squeeze the green color out.

And here is the second:

■ Lindsay, age 7, methodically pulls out strips of toilet paper of equal lengths and lays them across the width of the table. She then carefully pours dribbles of water on the strips. When she tries to gather the slightly wet paper up into a clump, she seems frustrated that it won't hold together, and begins to throw it down. The teacher, observing that Lindsay's frustration is preventing her from following through on experimentation, considers asking "What would happen if you added more water?" Instead, the teacher waits, continuing to observe. Lindsay tosses the paper into the water and begins to lay out more strips of toilet paper, placing them carefully into the water and watching them sink. She runs to get a sieve from the dish drainer next to the sink, comes back, and scoops the wet paper out of the water with the sieve. She then gathers the paper up in her hands and forms six balls, the same number as the number of strips she had laid out on the table.

SECOND ACTIVITY: "CHANGING CHAINS"

In the second activity—also designed from the question "How can I make it change?"—the focus is on experimentation with the rearrangement of the same original, recognizable parts (as opposed to the first activity, which focused on the changing of form).

Context and Brief Description of Activity   In this activity, children will be connecting paper "links" to each other and to various "hooks" placed nearby so as to create a multidimensional, changing sculpture.

Materials   Note that this activity, depending on the age of the child, will take some adult preparation time. However, the materials are quite strong and may be reused.

- large brown paper shopping bags (amount will vary depending on the number of children involved in the activity)
- construction paper (8½ by 11 inches) of different colors
- metal paper clips
- "Doughnut" hole reinforcers—one to two boxes
- string, yarn
- cup hooks, self-adhesive type, or any object (such as screen netting, old tennis rackets, etc.) that may be hung in the room and into which bent paper clips may be hooked
- scissors, tape, paper

Directions for Preparation of "Paper Link"   Cut paper bags or construction paper into rectangles of assorted dimensions ranging from approximately 3 to 5 inches wide and 7 to 12 inches long. Fold them widthwise beginning at one end and alternating the direction of the folds to create an accordion effect. Put a hole reinforcer at each end and poke an unfolded paper clip through the hole.* (See illustration.)

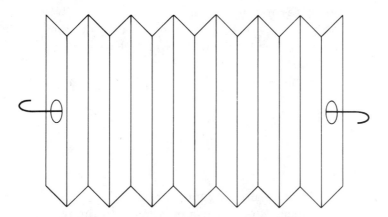

---

*Older children will be able to assist in this preparation, although this is not the focus of the activity.

Setting Up the Activity   Place at various levels and planes some of the hooks, loops of string, netting and so on in the section of the room where the activity will be. Some children will be interested in generating more possibilities for "hook" placement, particularly after the activity gets under way.

The activity should be set up in an area of the room where it can remain for several days. Some children might prefer to work on the changing chains for the entire activity period. Others might work on it and leave, letting other children enter the area to add extensions to the original structure and/or to change its form and direction. In this context, the teacher might be facilitating the children's additions by observing a child's experimentation and, for example, getting him or her a chair to stand on or looping string through a ceiling hook.

Ways to Introduce the Activity   As we discussed in the other activity introduction, the primary purpose of introducing an activity is to foster inquisitiveness and encourage children's experimentation. In order to plan an effective and appropriate activity introduction, you must think about the goals and rationales for your activity. For example, in order to facilitate children's awareness that they can create change by rearrangement of the same, original, recognizable units, you would want children to generate some different ways in which they might arrange the paper links and continue to change their arrangement. After showing the children the paper links and either pointing out the hooks or asking the children if they notice what is different about the room, you might ask them "What are some ways that you might hook these together to make a 'changing chain'?" After following a few verbal directions from some children, you might ask all of them "How can I change it now?"

There is not one "correct" way to introduce this activity. As before, the important thing to ask yourself when planning an activity introduction is, "What goals or rationale do I want to focus on?" The following scenario illustrates some possible responses children might make to this teacher's choice of activity introduction and questioning strategies.

■ To introduce the "changing chains" activity, the kindergarten teacher asks the children to seat themselves around her in the area where the activity will take place. Hooks, loops of string, and some nylon netting have been placed at various levels on some walls and shelves nearby. She asks the children, "What do you see that's different about this area of the room today?" Sierra responds quickly, "Those hanging hooks and the . . . " "There's a lot of hooks," interrupts Larry, "And look at the net hooked onto our block shelf!" The teacher then dramatically picks up a large paper bag filled with paper links, turns it over, and dumps it on the floor in front of her. All the children are very interested in the materials and listen intently when the teacher speaks and holds up a few of the links. "These are the paper links of our changing-chain activity. What are some ways I could put these links together to change how they are now?" A few children respond, "Hook them together!" "Like this?" asks the teacher as she connects three of the links in a line. Larry, who has been looking around the area at all the hooks, says, "Hook them to this hook." The teacher does this and then asks, "What are some different ways we could change our chains?" One child suggests hooking them straight down to the rug, another directs the teacher to add more links hanging down from the same hook. The teacher tries some of the suggestions as the children give her directions. After she has created a chain running horizontally to the floor about 3 feet above it, she asks, "Any predictions about what might happen if I changed this chain by unhooking one side of it from this hook on the wall?" "It would go straight down." "You could pick it up and move it up there." "It would be all wiggly then!" After unhooking one side and listening to the children's responses about their predictions and what they might do now, the teacher finishes the introduction by saying, "Here are a lot of the links to the changing chain. I'm interested in seeing all the different ways you can change the chain. Here's some more paper, pens, string, and tape you might want to use also."

How Children Might Engage in This Activity   Here are some possible ways children might engage in experimentation with this activity:

- Children might hook the links together in small sets of three to five and hang them from the various hooks.
- Children might connect a link to a hook and then begin adding links until the chain reached the floor.
- Children might connect a link to a hook, begin adding links until it was almost to the floor, and then change the chain's position to a horizontal one by connecting it to a hook in a different part of the wall.
- Children might begin connecting links from the horizontal chain to the floor.

- Children might add to other children's structures by making individual chains and then connecting them at various points on the hanging chain to create diagonally hanging loops.
- Children might change the structure of the chain by lifting it up at its midpoint and connecting it to another hook on the wall.
- Children might make more loops with string and tape them to serve as hooks at various places on shelves.
- Children might cut paper in different shapes, draw on it, and tape the shapes to the hanging chains.
- Children might unhook a section of the chain and drop it to the ground or hook it to another section of the wall or shelf to add another wing to the structure.

Theories, Questions, and Predictions Children Might Have While Engaging in This Activity   As stated previously, these "questions" children might be asking themselves through their experiments are not necessarily verbalized and are stated here in adult language.

- What if I hook links together? How will it change? How will it then change if I connect those links to the hook on the high shelf?
- How will it change if I keep adding links until it reaches the floor? How will it change if I add more and more after it reaches the floor?
- How can I make it change if I lift up the chain from the floor and connect it to a hook on the far wall? What would happen if I unhooked it from the hook on the wall and let it drop to the floor?
- What if I took some of the links off the end of the big chain, then hooked it to its middle, and then hooked that to the hook on the high shelf above me?
- What if I left the area and came back and the chain was all different? How could I change it and make it hang to the floor again?
- What if I cut some string and made loops to tape to the floor? What if I changed the direction of the hanging chain by hooking it to my loops on the floor?
- What if I cut some paper strips and taped them onto the chain that is hanging across the shelf? What would happen if I unhooked the chain from the shelf hook and added more links and connected the chain to a hook up high?

A Scenario

■ Stephen, age 5, and Anne, age 6, sit down on the floor in the area set up for the changing-chain activity. They both begin connecting links to form individual chains. After Anne has connected six, she stands up and looks around the area at the various hooks and decides to hang her chain from a hook on the high shelf. After looking at the chain hanging vertically, she gets some more links and adds them to the hanging chain until it is about 4 feet long. She then lifts up the end of the chain and stretches it horizontally in order to reach another hook. Stephen looks up from his work on his long chain and gets up and hooks his chain to Anne's chain. They both look at the greatly increased length and laugh. Ann picks up the unconnected end of the chain and hooks it to the other hook. The middle of the chain is now dangling on the floor and Stephen steps back and trips on it. Anne, frustrated, looks at Stephen and says, "You're going to break it . . . I don't want the chain there." The teacher, who has been observing and notices that Stephen is about to leave the area, asks the two children, "What are some ways you could change the chain so it would be in a different place?" Stephen carefully picks up the chain in the center and looks around the area. Anne exclaims, "I know!" and gets a piece of string and loops it around the center of the chain and directs Stephen to get some tape. Together the children tape the loop of string around the chain and then onto a chair which Stephen has moved near their work. They begin then to work on connected links to form individual short chains, which they hook to the larger chain until their weight causes the taped loop to come loose. The children look at the chain dangling on the floor and pick it up by the added short links and link them to some hooks on the shelf, changing the direction of the structure in still another way.

These sample activities and scenarios illustrate how the question "How can I make it change?" can be used to apply constructivism to the classroom. Transformation is such a focus of young children's activity and thinking that the possibilities are endless, both in what the teacher might plan and how the children will experiment with change.

Chapter
7

How Does It Fit or How Do I Fit? Constructivist Biology and Ecology

What is an appropriate focus for young children that incorporates some of the subject matter that we define as "biology?" Our focus, for reasons we will explore in a while, is on stimulating children's sense of wonder and appreciation of the natural world. In focusing on the development of the sense of wonder and respect for nature, we are probably placing more emphasis on ecology and what we call "ecological perspective taking."

For many early childhood educators, an emphasis on plants, animals, and ecology is the most natural and appropriate way to bring science into the classroom. In fact, for some of us, a "good" classroom *has* to include animals, bugs, and growing plants in order to be considered a warm and inviting environment for young children. And we can probably find more resources available for a curriculum about living things and on "nature" than we can in the areas of physics and chemistry (even as broadly as we have defined them).

A CONSTRUCTIVIST APPROACH

Our challenge in this book is not to talk merely about the incorporation of nature into the classroom. Rather, it is to try to define the *constructivist* approach to integrating nature into the classroom. And while we may not find it very difficult to think about ways of incorporating the natural world into our programs, it is a different story to ask ourselves what is constructivist about our work in this area.

Remember that constructivism focuses on the child's perspective, both on how the child sees the world and on what it is that the child is constructing. This *how* and this *what* are the two interacting components of the science curriculum—the *process* and the *content*. In this chapter, we shall be considering very carefully which elements of the natural sciences are congruent and would be emphasized in a constructivist early childhood classroom. We will be considering both the ways in which children learn about those elements (the process) and the particular contents that would be the focus.

Before we do that, let's consider why thinking "constructively" in the area of natural sciences is somewhat different than it is in the areas of chemistry and physics. The clearest problem we face is that it is harder to experiment in the natural world in ways that are appropriate for young children. In the other areas, we have been encouraging *experimentation* as the major way that children construct knowledge—experimentation that children engage in themselves, directly, with objects and materials. But the objects and materials in the natural world are often living things; while experimenting with them might be interesting, it is often not permissible. A child may ask "What will happen if I pull the tail off of the mouse?" It is simply not okay for the child to pull off the mouse's tail to answer that question, even if, in this way, the child could learn something about living things. Moreover, it is not only "not okay" but it would mean teaching something that undermines the ultimate goal of understanding the natural world. This goal means that as we learn about this world we develop respect and compassion for living things of all kinds—we refer to this as "ecological perspective taking."

This is not to say that in some ways experimentation cannot be encouraged; but in this domain of science, experimentation cannot be the *primary* emphasis. The alternative to experimentation is observation, which runs the risk of being too passive to be appropriate for young children. Yet observation can also mean sensitivity, alertness, and awareness of the natural world. This is, indeed, one of the keys to feeling a part of nature. As Rachel Carson talks about it, the "sense of wonder" that children naturally have and that we want to foster is based on just such an active observation of the natural world—observation that is unobtrusive and does not involve experimentation.

But it is not passive. We choose, instead of talking about just observation, to talk about participation. Participation implies that one is involved in some way with what one is observing. You can look at a spider weaving a

Careful planning and sensitivity to the animal are essential when children interact with living creatures. To ensure this mouse's safety, the children must constantly consider how their actions affect the mouse.

web, for example from a distance, objectively, without any sense of the spider as a living creature, without any connection between yourself and the spider. This is not our goal. Our goal is for children to participate with the spider as they watch the weaving of the web—for them to connect the spider's weaving with their own creation of space. In that way, children can begin to make the connections between living things that is the essence of ecology.

This may sound too mystical to be science. Yet such a view of biology and ecology is very consistent with current thinking and research—from work on the connections between rain forests, nonbiodegradable materials, and the ozone layer to theories about the earth as an organism.

How This Chapter Is Different

Such a shift in focus—from "experimentation on" to "participation with"—also entails some other different emphases in this chapter.

First, there will be less emphasis, in this domain of science, on planned activities (although there will be some) and more emphasis on creating an environment in the classroom and being prepared to capitalize on spontaneous events. This requires some different preparation on the part of the teacher, who must become better prepared to capitalize on spontaneous events with knowledge, resources, and interest.

Second, we shall shift from emphasizing children's direct actions on materials and objects to children's participation in and observation of the natural world. We'll discuss ways to bring this about in the context of the classroom that will not be overly passive.

Finally, there is the need for increased teacher guidance, with slightly more emphasis on the teacher as a model. Teachers will also have to ensure that children do not intrude on the natural world. Children will learn to love and know the natural world only through interaction with adults who show respect and love for it; we cannot assume that this will happen if we, as adults, do not serve as models.

The Teacher as Model

It is always important for teachers to model the interest in science that we want to encourage in children. When we focus on experimentation, this modeling takes the form of expressing puzzlement and curiosity, making predictions and offering hypotheses (some of which may turn out quite differently than we predicted!), and in general adopting an "experimental attitude." As we shift our focus away from direct experimentation, we still model a deep curiosity and interest—an ability to observe and "turn toward" interesting events, objects, and materials and an abiding respect for both living and nonliving nature. For some of us, this is a natural predisposition—you know if you are one of those who will wake up your children in the small hours to see a really remarkable moon, or who will drop everything, no matter how "important," in your second-grade classroom to look at the amazing spider web that a child has discovered. Others of us need to cultivate such a disposition. We can do this by letting ourselves focus on little things and by reading the wonderful accounts of nature that have been written for the "lay" person by naturalists such as Stephen Gould. One of the key books to help you cultivate and deepen your sensitivity to nature is Rachel Carson's *A Sense of Wonder*.

In addition to feeding your love for the natural world and awe at the myriad of amazing things we encounter daily, we can be better models if we feed our own knowledge base. Because you are an adult thinker, the more *you* know about butterflies, for example, the more you will start to notice many other things. The more you know about the rabbit in your classroom cage, the more you can serve as a resource by answering spontaneous questions and offering needed information.

Preparing yourself to serve as a resource when questions arise includes having access to resource books, materials, and people who can answer questions and give information when *you* can't. Becoming familiar with good references and community resources permits you to respond more quickly when the need arises.

Let's give an example of what we are talking about. How can you, as a teacher, prepare yourself for things you cannot predict?

■ Before the preschoolers had arrived the teacher discovered that Oatmeal, one of the school's pet mice, was dead. Wanting to share this with all the children together, she removed the dead mouse from the cage and put her in a box out of sight. Later that morning as the children are seated together for group time, the teacher announces, "Something happened and I need some help deciding what to do. Oatmeal died last night." Then she places the dead mouse on a piece of paper in the center of the circle. "Who made her dead?" pipes up Felicia. "Do you have some ideas about what might have happened?" responds the teacher. "Maybe a cat snuck in the crack under the door," answers Adam. Emily volunteers, "Maybe she's not dead. Maybe she's asleep." "Well . . . ," answers the teacher slowly as she strokes the dead mouse. "She feels real different. Would anyone like to feel her?" The teacher walks around the group of children slowly giving each of them who wants one a turn. The children are all very interested and very serious and their attitude reflects the teacher's "matter-of-fact" handling of the mouse. The children generate a lot of different theories about the coldness and stiffness of Oatmeal's body, and the teacher offers her knowledge about how a body changes when it dies. After a while the teacher repeats her original remark, "We need to decide what to do with her." The children all agree to bury the mouse. Because she had anticipated their suggestion and she wants to focus their attention on the process of decomposition, she has brought to group a box and a simple book about life, death, and the soil. Mark, 4 years old, announces that his book at home said that the mouse couldn't turn into dirt if it was in a box. "But," says Samantha, "I have a box in my yard that got stuck under my dirt pile and I found big mushy holes in it and lots of worms." The children finally decide to wrap the mouse in leaves and bury it in the ground and, in a different part of the yard, to bury a cardboard box, making predictions about what they might find when they dig up the box in a few weeks. Together they all go outside to bury the mouse and sing a song about flowers growing.

The teacher here was flexible enough, and prepared enough, to capitalize on an event she could not have predicted. In the way she handled the discussion and in the resources she brought to the situation, she was able to make this experience a meaningful and educational one for the children.

CATEGORIES OF BIOLOGY AND ECOLOGY

Because of their consistency with the emphasis we have chosen for a constructivist approach, the following three categories can help you to bring the natural world into your classroom and to develop activities.

The Perspective of Animals and Plants

Learning to live with animals and plants, understand their needs, respect them, and interact with them appropriately is a very natural way to promote ecological perspective taking. Children's natural affinity to and interest in living things is an instant entry point to learning about how other living things are both similar to and different from us.

But we are talking about going *beyond* just knowing the similarities and differences between ourselves and other living things. We want children to understand more than the dichotomies of "us" and "them." Instead, we want to communicate the complex interdependencies of the natural world through children's experiences with animals and plants.

This is parallel to the distinction made in Chapter 6, where we were talking about children understanding transformation. In that chapter we described how it was important to focus on the continuum of change instead of the beginning and end points. Similarly here, instead of focusing on A (the child) and B (the animal), we really want children to understand how A and B are mutually dependent and interacting.

Thus, while it may be useful and interesting to categorize the natural world into those aspects of animals and plants that have certain characteristics (animals that are meat eaters and those that are plant eaters, plants that are annuals and those that are perennials, bugs that fly and those that do not fly), we need to examine whether what we are pointing out will add to the child's understanding of the complex connections between animals and plants, between humans and other living things, between things that are animate and things that are inanimate. Sometimes this complex inter-dependence is best communicated by focusing on similarities and differences at the same time. For example, looking at the spider's web as a structure unique to the spider and concurrently at other creatures' struc-

Artificial flowers, potting soil, and garden gloves encourage symbolic play about planting.

tures would draw attention to the similarities among them as well as to the diverse array of structural habitats.

Bringing This Perspective into the Classroom   There are many different ways of communicating an understanding of the perspective of plants and animals. Encouraging dramatic play and facilitating role playing is one way. As we know, imaginative play gives children an opportunity to "step into the shoes," so to speak, of someone or something else and to test out these roles and other related content matter. Even fantasy themes that are "inauthentic" and far from reality provide opportunities for perspective taking. For example, the children might pretend to be spiders in a dramatic play area that is set up as a spider's web. In the classroom, you could create an area in which the children construct a web structure with room weavings. In building the "web" and playing in that area, children could start to see the stability of the web as well as its vulnerability to outside influences. The use of literature and pictures could connect the children's experiences in their self-created web with the webs that spiders build. The goal of all of these activities would be to encourage the perspective taking that can foster understanding.

Another tool for facilitating perspective taking is to have animals in the classroom, both as regular classroom pets and as visitors. Involving children in the care of animals, encouraging appropriate handling and playing with animals, and setting up activities that involve animals all can increase children's awareness of animals and their needs.

Let's see how that might happen.

■ One day the teacher announces to the class of first-graders that they will have a new pet in the classroom. Wanting the children to reflect on the needs of the animal, she suggests they play a guessing game. "Before we guess who our new pet is, you can ask me questions for clues." Immediately half a dozen hands shoot up. "Does it fly?" "Does it eat meat?" "Does it have bones?" "What color is it?" "Does it have fur?" "Does it like it to be hot?" "How fast can it move?" After answering all the questions, the teacher finally says, "Well, any predictions about what animal it might be?" "A salamander," shouts out Kevin. "No, it hops," reminds Jody, "Is it a frog?" "You guessed it," says the teacher. "Now that we know a little about the frog from *your* questions, here's *my* question. How do you think we treat the frog?" A lengthy discussion with a few pantomimed demonstrations follows.

Gary, who has a frog at home, volunteers to show the children exactly how to hold the frog when he brings it tomorrow. "I let my frog hop around at home," he says. "Could we do that here?" The children and teacher decide that the frog could hop around if they posted a list entitled, "How to be kind to the frog." The next day the children sit in a circle with their legs touching to create a barrier and watch Froggy hop around. Through discussion and continual embellishment to the "How to be kind to the frog" rules, the children are allowed to handle Froggy without asking each time and to let him hop around a designated area. The area where the frog can hop is eventually expanded as the children begin to create elaborate courses for Froggy and time his movement through and over their mazes.

The Sense of Wonder

The second category of curriculum that we have identified involves facilitating children's ability to focus on beautiful and interesting characteristics of the natural world. This is more than observation; it is providing opportunities for and setting up activities in which children can really "tune in" to

In the living world, experimentation takes on a particular meaning—gentleness is a necessity for developing an ecological perspective.

some aspect of the natural world that, while always there to observe, may be overlooked. Sometimes it is possible to be overwhelmed by the many things to observe; sometimes we really do not see the forest for the trees or the trees for the forest. Think of children out walking on a beach; there is so much to see, so much open space, so *many* interesting things that it is sometimes almost too exciting to be able to notice anything!

There are both tools and techniques for "tuning in" to the natural world. Tools are usually things that force us to focus on one aspect of the natural world.

1. *Magnifying tools*—a magnifying glass makes us zero in on a blade of grass, a rock, or a patch of dirt. A magnifying bug container makes us look intensely at one bug. A pair of binoculars lets us look at the bird perched among many birds on the fence. Little magnifying boxes that rocks can be put into permit closer scrutiny and comparison. The teacher may want to collect sets of rocks that can be scrutinized and compared.

2. *Capsulizing nature*—we are also more tuned in to parts of the natural world when they are encapsulated and placed where we can watch them over time. Small animals in cages, birds, bug houses,

ant farms, fish in aquariums, and butterfly gardens all permit us to study more closely, out of context, the structure and behavior of animals. They need not involve long-term imprisonment; butterfly gardens, for example, permit children to watch the progression of the butterfly and then release it to the wild. Appropriate uses of bug houses include short-term incarceration for the purpose of studying the bugs.

3. *Recording nature*—sometimes we "see" things in a photograph, film, or videotape that escape our eyes at first glance. (Videotaping the field trip on the beach and then using that as the basis for a discussion of all that was seen and done, or using a big poster blow-up of a bumblebee—these can focus our attention on details that might otherwise be overlooked.)

Here are two scenarios that describe how children can be encouraged to focus on characteristics of the natural world. The first is in a preschool.

■ Alonda is standing in the library looking out of the window. Suddenly she notices a dead fly on the windowsill. She runs over to the teacher, very excited, and exclaims, "There's a big bee in the window!" The teacher comes over to see, and, seeing that it is a fly says, "Great! What could we use to look at it closer!" Alonda pauses for a minute, goes to another part of the room, and gets the magnifying glass. Looking carefully at the fly, Alonda notices that it doesn't have a stinger. "It's a fly," she proclaims, "and I see some baby flies." Alonda points to some gnats caught in an old spider web. Together the teacher and Alonda look at the gnats and the fly, talking about their wings and bodies.

Here's an example from a second-grade classroom:

■ For the last four weeks the second graders have been taking walks over to the small grove of trees adjacent to the schoolyard. Each child has selected a tree to be "his" or "her" tree and they have familiarized themselves with the trees' characteristics. The children are keeping journals about their trees, using both words and pictures. Today the teacher has brought a video camera along. Each child has the opportunity to direct the teacher to different parts of the tree he or she wants videotaped. Upon returning to the classroom, the teacher plays the videotape and the children guess which tree is theirs. Pablo is sure that the first tree is "his." "See," he directs the children, "look at the bottom of the trunk, there's a scar line on my tree!" The teacher stops the tape while they examine it more closely. Betsy argues, "My tree has a line too, and look, the leaves are missing from one branch. It must be *my* tree." The children continue to examine the tape, sharing the attributes of their trees and deciding which tree "belongs" to whom.

Focusing on characteristics of the natural world can range from the spontaneous use of a simple magnifying glass to a planned use of more sophisticated equipment.

Transformation in the Natural World

As we have discussed in the previous chapter, transformation is a major focus of the child's world, and it is a particularly important part of nature. In fact, it is hard to think about studying biology and ecology without having transformation come into play. Birth, growth, death; decay and decomposition; changes in substances over time—all these transformations in the natural world can be emphasized in a constructivist classroom. As we discussed in the last chapter, we need to keep in mind those characteristics of activities and experiences that would encourage children to focus on transformation. Unlike the criteria for a good physical knowledge activity (see Chapter 5), which calls for the effects of the child's actions to be immediately observable to the child, and unlike the change activities in Chapter 6, in which the teacher plans for change which is immediate and visible, transformation in nature is usually less immediate and less visible to young children. In the activities and learning experiences we plan for, we must pay attention to the observability of the change processes—their accessibility to young children's thinking. This can be a challenge because, as we discussed in the previous chapter, transformation in the natural world does not lend itself to active experimentation by children. Also, in the natural world some of the most interesting transformations occur over long periods of time. For example, recall the description in Chapter 1 of the children who come across a decaying tree while they are out for a walk. How can we facilitate an understanding of such long-term change processes with which we cannot necessarily experiment in appropriate ways?

Let's look at some examples of how one might focus on transformation.

■ In a second-grade classroom, Susan has just returned from a camping trip. She has brought back a collection of rocks for the class, including some soft rocks such as pumice stone. "Look," she shows the class during group time, "when you rub them together, it turns into powder!" The class watches closely as powder falls to the floor. "Try it with these other rocks!" suggests Douglas, pointing to some hard rocks. Susan rubs two rocks together and no dust falls. The teacher asks the class, "Do you have any ideas about how we might change the hard rocks?" "With a hammer," suggests Linda. After a lengthy discussion about what might happen if bits of rocks flew up and hit someone's eye, the class decides to use goggles, hammers, and sand paper and to place the rocks in strong canvas bags while they pound on them. The teacher sets the activity up for the following day. The children work hard making rock chips and begin to bring in dirt clods, shells, and pieces of rotten wood to try to change by pounding on them.

This activity raises some interesting and difficult-to-answer questions about the line between experimentation and observation. One of our goals in the area of biology and ecology involves developing respect for and understanding of the natural world. Are we sending the wrong message to children by encouraging them to "smash" rocks, shells, and other things in order to explore their properties? At what point are the children's activities overstepping the boundary of showing that respect and appreciation for nature? Here is where the teacher would need to communicate to the children, probably in a group discussion, the issues involved. The teacher could involve the children in generating solutions to this "problem."

This difficulty doesn't always arise when we are considering transformation in the natural world. Let's look at two more examples that involve focusing on natural transformation.

■ After the preschool children begin to tire of playing outside in the snow, the teacher decides to bring some freshly fallen snow inside. She places the snow in a small swimming pool with some pans of various sizes. Cooper begins to pile the snow up high in the containers and takes them over to a table by the heater to "cook." Trish watches him set the dishes down and asks, "Want to come have snack with me?" Cooper glances at his containers and follows Trish to the snack area. About halfway through their snack, Cooper jumps up and announces to Trish, "I'm going to check my food!" He goes over to the table, where he presses down on his piles of melting snow. He slaps the water with his hand, picks up some of the still-solid snow, and squeezes it together to form a hard ball. Cooper then returns to the snack table to offer it to Trish, saying, "Here's some of my snow candy, the juicy part is all over the stove!"

In the above example, bringing the snow inside speeds up a process that occurs naturally outside.

■ The kindergartners were excitedly crowded around the mice cage to see the litter of mice that Chocolate had just given birth to. Rachel questions the teacher, "Why are they pink? The mom is dark brown." Rob announces with an air of authority, "They change. You watch and see!" The teacher asks the children if they have any predictions about how the baby mice might change. "They'll get bigger!" "They'll turn white," and, "Their fur will grow," were some of the suggestions. The teacher gets out her instant camera and asks if the children would like to take pictures of the baby mice each day to help watch how they might change. This daily picture taking becomes an important "ritual" for the next few weeks. The pictures are placed on a bulletin board and the children supplement the pictures with dictations and drawings about the mice.

The pictures help make the processes of change more observable to these children.

In the following section, we will describe in more detail two learning situations that embody a constructivist approach to biology and ecology. The first situation arose spontaneously and was extended by the teacher and the children, while the second was planned by the teacher. Depending on the geographical location of the school and the accessibility of the outdoors, teachers will have varied balances of spontaneous and planned activities. Sometimes the curricular format will go from spontaneous learning situations to teacher-planned activities on the basis of child observation and a knowledge of the subject matter. Other times the children will begin engaging in a teacher-planned activity and extensions will arise spontaneously, facilitated by the teacher. Let's look at one spontaneous event and discuss how the teacher created an environment to extend the children's awareness.

Spontaneous Learning Situation: Spiders

■ First-grader Laura is carefully and proudly doing her "helping hand" task of watering the three plants on the windowsill. All of a sudden she exclaims, "Something new came on the big plant today!" Several nearby children come over to see and, as the teacher looks up from across the room, she hears Mike shout, "Hey, a big fat spider! and it is making a web!" The children who came over back away a little nervously. At this point the teacher comes over and notices some of the children's apprehension.

How the Teacher Responded to Enhance Ecological Perspective Taking
The teacher's goal is to replace some of the children's apprehension with respect for the spider and for the children to notice some things about the spider. The teacher might choose to focus on the spider's web or on the fullness of the spider's body (the teacher is guessing that the spider is full of eggs). Depending on how observant or rambunctious the children are, the teacher might choose to focus on how the children's actions affect the spider, or she might focus on what they can see if they look carefully. The teacher could make comments or ask questions such as, "I wonder what made the spider so fat?" "Do you have some ideas about what the web is for?" "I wonder how it makes the spider feel when you wave your arms around so near it."

■ Laura and Mike express concern for the spider and worry that someone might bump the plant and break the web. The teacher asks them how they could tell the other children to be careful. The children decide to make a sign and busy themselves with this task. They also ask the teacher if they could talk about it at group time. Meanwhile, other children come over to see what is going on. Anticipating the appearance of baby spiders in a few days, the teacher hopes the children will all notice the size of the spider's body. So that they will, she sets some magnifying glasses nearby, which the children immediately begin using to examine both the spider and the web.

Themes for the Group Discussion That Day   Before group time following this incident, the teacher thinks carefully about discussion points that might enhance the spontaneous spider experience. She prepares herself by selecting a well-illustrated book on spiders. The teacher is also aware that the discussion might take off in a different direction, depending on what the children noticed and also what personal knowledge and theories they are bringing to the experience. The teacher hopes to discuss with the children one or more of the following topics: how the spider web might change, what the spider web is for, how to act around the spider, and what might be making the spider's body so big. Underlying these questions are the concepts of transformation, the similarities between people and spiders (the needs of spiders and people for food, for example), and a sense of wonder.

Here is how the discussion actually unfolded:

■ First, Laura and Mike tell the group to be very careful not to scare the spider because it might run away. Laura also adds that if the children wave their arms it might break the web. Another child says that the web is strong and won't break, just bend, and if it does break the spider could just build another. Then the group starts making predictions about what might happen and talking about how the spider makes the web. After listening to their theories, the teacher reads out loud to the group a section in the book about spiderwebs. Suddenly, James blurts out that his Aunt Joanne got bitten by a brown spider and that it made a hard, sore place on her arm that is not going away, and that she got bitten when she was outside working in the garden. Since the bell for recess is about to ring, the teacher asks the children if they have some ideas about what they might look for when they are outside today. James says, "Spiders and webs, but be careful not to get bit like my Aunt Joanne!"

Here is what happens during recess the first day:

■ The teacher brings out some magnifying glasses for the children to use. The children find some spiders in the grass and some old abandoned webs. The teacher goes into the classroom and brings back outside some black paper on which to "capture" the empty webs and has the children help her sprinkle them with white flour. Because the flour sticks where the web is, it makes a "web print." She makes sure that the children know that the webs are no longer being used by the spiders. The teacher also uses her camera with a closeup lens to take some pictures of the spiders in the grass. During the spider search, one of the children finds an abandoned bird's nest. The children bring it into the classroom.

Emerging Themes or Topics   On the basis of what happened at recess, the teacher sees that some topics are emerging from the children's actions. One is animals' need for "homes" and their reasons for leaving their homes. She plans to help the children connect these interests with their own need for a home. She also hopes to tie this in with the upcoming move of one of the children to a different neighborhood. Another emerging theme, based on the fact that the children noticed bugs caught in the webs, is the need for food. And still prevalent is the concept that we must respect the "space" of other living things, for their sake as well as for our safety.

Here is how the teacher facilitates the development of these topics:

■ The children place the web prints around the room and, as soon as the film is developed, the pictures are glued to a large poster. Some of the children dictate their ideas about the spiders to the teacher, and these are put on the poster. The children are asked to bring pictures of their homes or rooms to school, and these are put up too. The children engage in a lot of group discussion about where animals live and what they eat, and the teacher takes a group dictation about their ideas. She also finds more books on these topics. When the children notice a bug caught in the web in the classroom the next day, the teacher creates a chart system so that the children can graph the number of bugs the spider catches. One day, the teacher sets up the room for room weaving so that the children can make a large "spider-web" of yarn and string that they drape across a corner of the room and attach to hooks, weaving more string in and out of the attached string. Some of the children begin playing "spider and bug" and eventually put on a little play for the class. On another day, James finds a spider loose in the room and expresses his concern to the teacher that someone might get bitten as his Aunt Joanne did. The teacher shows the children how they can carefully and safely capture the spider in a cup (with a stiff piece of paper slipped under the inverted cup) and take the spider outside. Finally one day, as anticipated by the teacher but not by the children, hundreds of baby spiders appear by the plant. The group solves the problem of what to do with the babies and how to take them outside. Thus emerges another theme, birth and infant forms of life. This theme continues until the end of the year, complemented by the mice's birth, field trips to pet shops and nurseries, and the birth of Laura's baby brother.

Thus, through a clear vision of the goals of ecological perspective taking, by being sensitive to the emerging interests of the children, by developing her own knowledge base and planning well, and by asking as many open-ended questions as possible, the teacher was able to create numerous valuable and cohesive learning situations from Laura's initial observation of the spider in the plant.

Planned Activity: Soil Exploration

Next, let's look at how a kindergarten teacher might *plan* an activity using the same constructivist approach, and what extensions might come from it.

Because it was autumn and many leaves were falling and would soon be decomposing into soil, the teacher wanted to focus on soil—different types of soil and what is in it. He knew that he should try to find soil that the children could actually see different things in, and he wanted to focus on transformation in nature. To begin with, the teacher asked each child to bring a pint container of soil from either home or some place he or she had visited. He planned his requests so that, each day for a week, five children would bring their soil sample in. The plan was to create an air of interest and anticipation about the soil samples and to take the time to look at everyone's samples during group time. At the end of the week, the teacher set up a "soil center" in the following way.

Materials
- long narrow table (so that the children would go down the table and take their time exploring before they mixed all the soils together)
- soil samples
- blank white paper next to each of the samples
- sieves with various sizes of mesh
- spoons and stirring sticks
- empty containers
- magnifying glasses

Context of the Activity   This activity was set up as a choice where children could come and go and spend as little or as much time as they liked during free-choice time. The teacher set up this activity away from accessibility to water because he did not want water to interfere with the children's ability to sift out things from the soil. He planned on adding water at a later date.

Ways to Introduce the Activity   As we discussed in Chapters 5 and 6, an activity introduction should be tied to the goals of the activity and should serve to encourage experimentation and exploration. Because the teacher wanted to focus on different types of soils, what is in soil, and how these things might change, he planned to introduce the activity by asking the children, "Do you have any predictions about what we might find in the soils?" and "How have these things changed?" The following is an example of how one activity introduction might unfold.

■ The teacher places a large piece of white butcher paper in front of him on the floor. On the floor he dramatically dumps three piles of different soil samples he had collected. Without saying a word, he scoops some soil in his hand and lets it drop back on the paper. "Any ideas about what I could use this for?" he asks the group while holding up a sieve. "Put some dirt in it and shake it!" Paul suggests. The teacher purposefully puts some dirt in and shakes it so lightly that nothing comes through and says, "Like this?" The children laugh and tell him to do it harder. The teacher does and some fine soil comes through, leaving larger pebbles and leaves in the sieve. "Do you have any predictions about what is in here?" he asks this time. The children suggest things such as "big pieces of dirt," "rocks," and "a rusty nail." The teacher then pulls out what is left for the children to see. He pulls out a partially decomposed leaf. "A leaf!" exclaim several children. The teacher then asks the children how this leaf has changed from when it was on the tree. "It's real holey," "You can crumble it now," and "It's getting skinnier," were some of their responses. The teacher announces that the project table will have all their soil samples on it and that there will be sieves and magnifying glasses so that they can find out about all the changed things in the soil.

Ways the Children Might Interact with the Materials

- Children might walk along the length of the table feeling each of the samples.
- Children might put some soil in the sieve and sift out some of the larger parts of the soil.
- Children might add some of these parts to other soil samples.
- Children might look at different kinds of soil on the paper with a magnifying glass.
- Children might mix the soils together.
- Children might spontaneously sort what they find in the samples, (for instance, rocks, bugs, leaves).

Possible Questions or Theories the Children Might Be Exploring During This Activity

- How do the textures of the various samples vary?

- What are some different things that I can find in the soil?
- Can I find the same things in soils that look different?
- How can I change the soils by adding different things that I sifted out?
- What can I see with the magnifying glass that did not come out in the sieve?
- How can I change soils by mixing them together?
- How do things change when they are in the soil?

Here is a sample scenario of children engaging in this activity:

■ Vanessa and Shelley approach the project table and walk all around it poking their fingers in each of the samples and giggling. Vanessa says, "Let's make leaf pie." Shelley says, "But I don't have any leaves." Vanessa directs Shelley to feel all the samples for leaves. The two girls busy themselves making a pile of soil they are sure has no leaves in it, sifting out leaves and other big clumps. Shelley goes to the drama corner and gets a pie pan and brings it back for their "pie." "First put the rocks on the bottom and then the big leaves and then the little pieces of changed leaves, okay?" suggests Vanessa. "I want to put crumbly pieces of wood in the pie too," Shelley informs Vanessa. The two girls continue to sift and sort the various parts of the soil, all the while discussing what ingredients should go into their pie.

Extensions from This Activity   With this activity the teacher intended to begin focusing on what is in soil and how different things fall into the soil and change. After doing the activity, he took the children on a hike around the neighborhood to collect soil samples under trees where the leaves had fallen. The children also collected leaves. One of the children suggested that they could "make soil," and the class collected more items from off the ground outside for this project. The teacher provided hammers, goggles, scissors, and grinders for this activity. He also found a book about "the decomposers"—the bacteria that decompose soil—and a child who walks through the woods wondering what happens to the leaves after they fall. The children enjoyed this story and acted it out over and over, using collected leaves as props. One day the teacher asked the children how they could make "magic soil" and provided wands, caps, and magic hats for them to use while they combined their suggested ingredients, such as soap flakes, dirt, flour, and clay. That day one of the children requested water for the soil mixture. This inspired days of exploration with adding water to real soil, both clay-like and sandy, to make paste and pliable clumps. The children experimented with the transformation from dry to wet and wet to dry using the south-facing window and heater as facilitators of the change process. Eventually, the children made a compost pile and in the spring were able to use this to plant their garden.

Whether the activity is initiated spontaneously or planned by the teacher, it is fundamental to a constructivist approach to ecology and biology that further activities and directions of exploration emerge from the children under the observant and responsive eye of the teacher.

FOUR

Final Thoughts

Chapter 8

The Teacher as Theory Builder

Throughout this book, we have focused on a constructivist educational environment that supports the child as a scientist and acknowledges the child as a theory builder. We have placed a high value on children's experimentation. We have shared both the methods and theory underlying our methods for encouraging children to be reflective. We have discussed how we can facilitate children's experiences as they seek to construct relationships and to consider different perspectives. And we have placed a high value on social interaction in contributing to children's development and their theory building. In orchestrating an environment that supports these goals, the teacher must assume many different roles, as we have discussed. When we analyzed these roles, we realized that the same goals we are promoting for young children also apply to teachers.

If teachers are to teach science (or any domain) from a constructivist perspective, they too must be theory builders. A teacher must have the courage to experiment. A teacher must be reflective, seeking relationships between the different variables in the classroom and considering different perspectives. And sometimes, just like children, teachers can be supported in their theory building by other teachers in a social context.

In thinking about the teacher as a theory builder, it is helpful to keep in mind two different categories of teachers. One type would be the "teacher in transition"—one who is seeking to implement a constructivist approach to science education for the very first time. This would also include the new teacher. The other type of teacher is the one who has already

The teacher must also be a theory builder, continually reflecting on the environment and activities.

begun to teach constructively and is seeking to implement a constructivist approach to an even greater degree. Most of the following recommendations apply to both types, although some of the ideas have particular relevance for the teacher in transition who may need more direction and support in being a courageous theory builder.

- *Question everything you do.* Don't make assumptions. Don't assume that what you planned to happen automatically *will* happen.
- *Take risks—experiment.* Look at experimentation (even "failure") as a way to learn and to grow.
- *Be autonomous.* Teach in a way that fits in well with your style and your classroom.
- *Don't expect change to be total and immediate.* Feel free to make incremental changes in your teaching style and environment.
- *Transform your environment and activities on an ongoing basis.* A *large* part of a constructivist science curriculum will come from the children themselves. The emergent nature of the curriculum requires the continual observation of children and development of new activities and approaches. Don't get into a rut by merely repeating your successes.

- *Seek information, new understandings, and new approaches through professional development.* Read books, take classes, go to workshops, get involved in professional organizations, and observe other programs.
- *Establish a support system for yourself.* Support systems can facilitate the exchange of knowledge and provide encouragement.

Now let's look at these recommendations in more detail.

QUESTION EVERYTHING YOU DO

All teachers, no matter what their theoretical perspective, should look critically at everything that goes on in their classrooms. This is particularly important for the constructivist teacher, who seeks to honor children's theory building and to constantly evaluate the classroom environment. Ask yourself, "Is what I anticipated actually happening? If not, what is? Is it appropriate for my philosophical perspective, my educational goals, and the children?" For teachers in transition who are seeking to implement a constructivist approach for the first time, much of this self-evaluation will include looking critically at some of their current beliefs about teaching. Constructivist teachers must be observant about what it means to be constructivist. They must also be reflective about the relationship between their rationale and what actually occurs.

Let's consider some different situations in the classroom and see how this need to question and reflect might apply. Here is the first:

■ A first-grade teacher who has been teaching in a very traditional, teacher-directed manner would like to begin changing the physical environment of her classroom. She decides to make certain items such as glue, scissors, staplers, different types of tape, and paper accessible to the children in order to encourage them to act on their own without waiting for the teacher's permission. She creates a low storage shelf against the wall at the back of the classroom. However, the children have not begun to use the supplies. The teacher then decides to lead a group discussion with the children to make them aware of all the different materials freely available to them. During the next few days, the teacher notices that only a few children are using the supplies. Upon observation and analysis, the teacher realizes that only the children seated near the supply shelf are using the supplies. It appears that children in the other parts of the room do not have a clear, easy route or pathway to use to get to the supplies. Additionally, after the teacher stoops down to "child level," she realizes that the view of the shelf is obstructed by a higher shelf, so that many of the children cannot even see the storage area. That day, after school, the teacher rearranges the room, placing the storage supplies in its center so that all the children will have both visual and physical access to the supplies. As a result of the teacher's analysis of the situation and subsequent change, the children spontaneously begin to use the supplies, incorporating more experimentation and creativity into the classroom environment.

Here is another:

■ A preschool teacher sets up a maze that is suspended from the ceiling so that the children can stand around it, tilting it to direct the movement of a ball. The teacher is very excited about implementing this constructivist activity in her classroom. The children, however, are not using the maze as a maze. At first the teacher is disappointed at the children's "fooling around" and feels that the activity is not successful. However, after observing the children's interactions with the suspended maze, she realizes that the children are busy experimenting with swinging the maze back and forth and around like a pendulum. She leaves the maze up and creates a pendulum activity in another part of the room. Eventually the children gravitate to the pendulum activity for their experimentation with swinging and some of the children return to the maze to experiment with tilting the maze in order to move the ball.

And here is a third:

■ A kindergarten teacher creates a painting activity to encourage children to work together and to consider each other's perspectives. The teacher attaches five paintbrushes perpendicular to a 4-foot board and attaches handles to each end of the board. She anticipates that two children will paint with this large "brush," each holding one end of it. However, the children choose to work alone with the brush, rather than in pairs. The teacher analyzes the activity and realizes that the brush, although large, is light enough for one child to handle alone. The material does not "require" that two children work together in order to use it. After the teacher adds a weight on one end of the board, thus making it difficult for one child to handle alone, the children begin to use the brush in pairs. By making a slight alteration, the teacher is able to create an activity that requires children to work together and coordinate their perspectives, as she had intended.

In all three of these situations, observations and reflection yielded significant results. Had the teachers not looked critically at their plans, they probably would either have been merely disappointed with the results or mistakenly assumed everything was going as anticipated.

TAKE RISKS—EXPERIMENT

A natural partner to questioning and self-reflection in the classroom is allowing yourself to take risks and to experiment. The real key to feeling truly free to take risks is to view each new plan, activity, or interaction as an opportunity to learn. We know as constructivist teachers that we must provide opportunities for children to experiment and learn through their "failures." We know, for example, that the child who tries unsuccessfully to tape a heavy block to a vertical surface is learning quite a bit! This same approach applies to you as a constructivist teacher. Sometimes you may have an idea for an activity but might not be sure if it will work out or be appropriate. Plan it carefully and view it as a wonderful opportunity to observe, learn more about children, and see the relationship between the materials and the children's interactions with them. This attitude of experimentation adds a creative dimension to teaching and learning and contributes a dynamic attitude to the classroom. The following examples will illustrate how this might occur.

■ The preschool classroom has a two-level wooden cube. The bottom level has archways on two of the sides so that the children can easily enter and the top level has a railing all around it. For the archway, the teacher constructed a "drawbridge"-type door, which is raised and lowered by rope from the top level of the cube. One day, she notices children putting beads on the drawbridge from the top level and watching the beads slide down the drawbridge. She decides to make another drawbridge of clear plastic sheeting, hoping that the transparent quality of the plastic will encourage the children to watch the objects slide and view their movement from both above and below.

Some different outcomes might result from children's interactions with this new "drawbridge." They might use it as the teacher anticipates, or they might use it very differently, releasing the ropes all the way so that it becomes a vertical "window" through which children could make faces at each other. Or perhaps the children won't use it at all. If that happens, the teacher can observe the dynamics of the day in order to understand why the new material is not being used. The reason for its lack of use might be as simple as the distraction of another exciting activity in another part of the room. Or perhaps the drawbridge has some technical flaw so that it does not function easily enough. Regardless of what actually happens, the teacher's new ideas, whether they yield the anticipated results or something different, are an important part of a constructivist teaching style.

Sometimes, the children can be involved in considering the feasibility of new ideas. For example, suppose a teacher has an idea for a new pulley system that will allow children to send messages across the room to each other. The teacher is not sure if this system will be in the way of other activities. She might present her idea to the children, showing them the hardware and rope and her plan, or perhaps she will ask the children where they think it should go. After the pulley system is set up, the class can evaluate it and collectively decide if or what alterations are necessary. In this way, the teacher is able to be an experimenter, involving the children in the issues of experimentation and encouraging their reflection.

Administrators or other observers can also be involved in the consideration of new ideas. For example, a first-grade teacher rearranges a section of her room, placing certain art supplies near the literacy or writing area. She does this to encourage reciprocity between learning areas, hoping to involve children in all facets of making books. In doing this, she also hopes to encourage those children to write who do not usually gravitate to the writing area. When the principal comes to observe, the teacher explains the change, her rationale, and her anticipated results. She also explains that she will be observing to see if these areas are compatible. Now the principal is aware of her creative endeavors and her reflective teaching and may become interested in seeing how the children use these areas. The teacher has involved the principal in this constructivist approach to teaching.

BE AUTONOMOUS

Constructivist classrooms all look very different. They *should* look different, not only because they need to reflect the individuality of the children in the classroom but also because the classroom is a reflection of the teacher's individuality. Teachers have many different inner reasons for creating different learning environments. They have different tolerances for noise, for messes, and for flexibility in both scheduling and the physical environment. Teachers might also have different *outer* reasons for creating their learning environment. For example, teachers might need to consider different expectations from the administration, parent groups, or colleagues with whom they share space or resources. The issue of "messes" seems to be a common concern and serves to illustrate the need for teacher autonomy. For example, if three teachers visit a "model" constructivist kindergarten program at a university laboratory school, they may observe a science activity in which children are experimenting with properties of liquids and solids by mixing together water, flour, salt, and food coloring in various combinations. Some of the children take sand from the sand table to add to their mixtures. Others decide to add some tissue paper from the supply shelf. Water and sand spill on the floor, but the children, with the teacher's help, sweep and mop up the floor. All three observing teachers are impressed with the children's interactions with each other and the materials and with their focused play. However, each teacher chooses to implement this activity differently in her own classroom. Let's see how different degrees of autonomy are apparent in their different implementations and attitudes.

The first teacher wants very much to implement a constructivist approach to science in her first-grade classroom. She does not, however, feel comfortable with the degree of mess the children make at the laboratory school. Feeling that she must "exactly" recreate the activity, she sets up the activity as she observed it. After one day of mess, she decides that she just is not cut out to be a constructivist teacher. This teacher was *not* autonomous in her implementation of the activity.

The second teacher, who also teaches first grade, had the same immediate reaction about her tolerance for mess. This teacher, however, considers her attitude and acts autonomously in implementing the activity. She decides to create a smaller-scale version of the same activity—using smaller containers, all placed in metal trays—and she limits the number of children allowed at the area. This implementation is constructivist, and it fits her own internal restraint about messes. Observing interesting learning encounters and happy with the results, this teacher gradually begins to branch out to increase the freedom she gives the children.

The third teacher works at a preschool where she shares space in a church building and must take down her program each day. This teacher's immediate feeling is that she won't be able to implement such an activity. However, because she sees the value in it, she decides that it is worthwhile to make arrangements to use an outside area for a large-scale, messy activity.

Sometimes autonomy will take a different form. Suppose you were in a school where you were the only teacher interested in teaching constructively. Sometimes autonomy calls for courage and assertiveness. As we discussed in Chapter 4, the teacher as a "public relations person" must be proactive, well informed, and articulate in order to be autonomous and to implement a preferred approach.

The important thing is that autonomy is just as important for constructivist teachers as it is for children in a constructivist classroom. Whether you are a "teacher in transition" or a teacher seeking to become increasingly constructivist, your theory building and experimentation must take place within the context of autonomy.

DON'T EXPECT CHANGE TO BE TOTAL AND IMMEDIATE

This applies particularly to the "teacher in transition" or beginning teacher. Often the changes that come about gradually are the most successful and are most likely to result in a true grasp of constructivism, with long-lasting results. Conversely, expectations of overnight change and success can often lead to disappointment and subsequent retreat back to "old" teaching styles. A teacher in transition might be most comfortable with first changing one area of the physical environment such as adding a shelf for supplies that is easily accessible to the children. Or perhaps a teacher might choose to begin with one constructivist activity, moving desks clear of one area of the room so as to set up a marble rollway activity with cut tubes, blocks, tapes, and marbles. Another way to move into constructivism gradually is to begin by changing only one area of the curriculum—such as science—starting with this book.

However, even though gradual change might be more comfortable for the teacher in transition, it is important to understand that the attitudes so essential for children in a constructivist environment—being autonomous and willing to experiment—do not exist in a vacuum. For example, it might very well be difficult for children to feel like experimenting in science when the rest of their learning activities are rigidly prescribed.

TRANSFORM ON AN ONGOING BASIS

The flip side of not feeling pressure to change overnight is the importance of continuing to change and to grow. The exciting thing about a constructivist approach to science education—or constructivism in any curriculum area—is that the children can give you continual inspiration for new ideas. In fact, observing the children for curriculum development is the backbone of a constructivist approach, and the constructivist teacher must be responsive to children's wonderful ideas.

Continual transformation does not necessarily imply that everything you do is always different. Let's take some basic materials such as tubes, blocks, and balls. These same materials might be set out every day, with children continually adding variations and increasing levels of sophistication to the use of these materials. They might begin with using them propped up against chairs for simple incline play, progress to making elevated inclines across shelves, and then begin to direct the balls' movement to targets. Or perhaps the children might use these materials in one way over several days and later not even seem to notice the tubes and balls. At this point the teacher might choose to put these materials away and focus on something totally different, or she might choose to introduce another material—such as tape or loops on pulleys to put the tubes through—in order to extend the children's learning encounters. Whether change comes about as a result of the children's spontaneous actions or whether it is introduced by the teacher as a result of her observations, transforming the activities and/or environment with good cause is an integral part of the constructivist classroom.

SEEK NEW INFORMATION/KNOWLEDGE THROUGH PROFESSIONAL DEVELOPMENT

The concept of transformation applies not only to what you do in the classroom but also to your own intellectual growth. C. T. Fosnot (1989) describes processes of teacher education from a constructivist perspective and suggests that we must reconceptualize teacher education to be consistent with constructivist principles. But regardless of your educational background, many formal avenues of inspiration and change are available. The consideration of new ideas and theories of curriculum and development can serve to keep your teaching vibrant. Professional organizations such as the National Association for the Education of Young Children and the Association for Childhood Education International have journals and newsletters published on a regular basis. Many of these organizations hold national and statewide conferences where teachers can attend various workshops. Such conferences can also provide exposure to displays of new materials and offer access to other teachers who are trying to grow. Classes at colleges and universities can provide similar opportunities and can also, through degree programs, add credibility to your approach. Books, both applied and theoretical, are a good way to gain new ideas and understandings. And visits to other classrooms can stimulate new ideas and self-reflection.

ESTABLISH A SUPPORT SYSTEM

Sometimes children learn best in a social context where they have to communicate their ideas to others and work together to achieve common goals.

Children also need support—both intellectual and emotional—from adults for their experimentation and for their theory building. Teachers—both those in transition and those seeking to become increasingly constructivist—have to create support systems among their peers, since it sometimes takes courage to give up old ideas and seek change. This support might simply mean having another teacher you can call on and analyze a particular activity with, to help figure out what went wrong, or to share your excitement. This support might become more formal if you meet regularly with another teacher or a group of teachers to share ideas, problems, and encouragement. Often formal organizations can be of great help. Another person can see causes or reasons for successes and failures that are not obvious to you. However you choose to establish support for yourself, it is essential for both your emotional health and your career as a teacher that you have someone you can turn to for encouragement and stimulation.

SO IS THIS REALLY SCIENCE?

There are those who might argue that what we have described in this book is not *really* science. There are those who might contend that it is stretching things to say that physics for young children "only" has to do with acting on objects; that chemistry is transformation; and that biology can be as "simple" as the child's "fitting in" to the natural world. To those arguments we say: Look at the young child. Think about how new it all is, think about how much he or she must comprehend. And then think about how complicated it all really is. What we have tried to do is think about what is appropriately comprehensible to the young child who still has that naive and yet driven curiosity to figure it all out.

There are those who might argue that we don't need to talk about all this as science at all. Just talk about the constructivist curriculum. And why try to redefine science anyway?

To those arguments we would say, again, look at young children as they engage in the activities we describe. They *are* scientists. They truly *are* actively experimenting, building their own increasingly complex theories about the world and the way it works. This joyful, engrossing, and ultimately constructive activity is what we want science to be for everyone. And this view of science as exciting and joyful should "trickle up" to intermediate and secondary education rather than being stifled at the lower levels.

As Piaget says, "The primary goal of education is to create people who are capable of doing new things, not simply of repeating what other generations have done—people who are creative, inventive, and discoverers."

The young scientist—a joy to behold!

References

Carson, R. (1968). *The sense of wonder*. New York: Harper & Row.

DeVries, R., and Kohlberg, L. (1987). Programs of early education: A constructivist view. New York: Longmans.

Driver, R. (1983). *The pupil as scientist?* New York: Taylor and Francis.

Forman, G. E., & Hill, D. F. (1984). *Constructive play: Applying Piaget to the preschool*. Reading, MA: Addison-Wesley.

Forman, G. E., & Kuschner, D. S. (1984). *The child's construction of knowledge: Piaget for teaching children*. Washington, DC: National Association for the Education of Young Children.

Fosnot, C. T. (1989). *Enquiring teachers, enquiring learners: A constructivist approach for teaching*. New York: Teachers College Press.

Gamberg, R., Kwak, W., Hutchings, M., & Altheim, J., with Edwards, G. (1988). *Learning and loving it: Theme studies in the classroom*. Portsmouth, NH: Heinemann.

Kamii, C. K. (1985). *Young children reinvent arithmetic: Implications of Piaget's theory*. New York: Teachers College Press.

Kamii, C., & DeVries, R. (1978). *Physical knowledge in preschool education*. Englewood Cliffs, NJ: Prentice-Hall.

Katz, L. G., & Chard, S. (1989). *Engaging children's minds: The project approach*. Norwood, NJ: Ablex Publishing Corp.

Osborne, R., & Freyberg, P. (1985). *Learning in science: The implications of children's science*. Portsmouth, NH: Heinemann.

Pappas, C., Kiefer, B., & Levstik, L. (1990). *An integrated language perspective in the elementary school*. New York: Longman.

Perret-Clermont, A.-N. (1980). *Social interaction and cognitive development in children*. New York: Academic Press.

Piaget, J. (1970). Piaget's theory. In P. Mussen (Ed.), *Carmichael's manual of child psychology* (3rd ed., vol. 1). New York: Wiley.

Sylvia, K., Roy, C. & Painter, M. (1980). *Childwatching at playgroup and nursery school*. Ypsilanti, MI: The High/Scope Press.

Appendix: Materials for a Constructivist Classroom

Chapter 3: General Materials

Plexiglas
    Tubes
    Small sheets
    One large framed sheet
Tubing
    Plastic, both clear and opaque
    Cardboard
Containers, including boxes
Play dough
Sand/water tubs
Blocks
Pulleys and hooks

Chapter 5: Materials Specific to Movement

Commercial products
    Marble rollways
    Gears
Balls of varying sizes, densities, and surface materials

Chapter 6: Materials Specific to Change

Commercial products
    Lego and Duplo sets

Tinkertoys
Mobilos
Flexi-blocks
Googleplex
Ramagon
Popoids
Play-Doh
Large sheets of vinyl
Art supplies, including powdered and liquid tempera
Flour, salt, cornstarch, shaving cream, soap flakes
Sand
Food coloring
Glitter
Clay of different kinds
Paper—all types
Glue
String
Clear containers, sieves, funnels
Paintbrushes and rollers
Medicine droppers

Chapter 7: Materials Specific to Natural World

Magnifying tools
"Encapsulators": cages, bug catchers or houses, ant farms, aquariums
Cameras, videotape
Resource books

Index

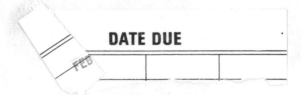

DATE DUE